Physics by Example contains two hundred problems from a wide range of key topics, along with detailed, step-by-step solutions. Applying the knowledge gained from lectures and textbooks to actual problem-solving is not always straightforward, but by guiding the reader through carefully chosen examples, this book will help to develop skill in manipulating physical concepts.

The book deals with the following areas: dimensions, errors and statistical analysis; classical mechanics and dynamics; gravitation and orbits; special relativity; quantum, atomic and nuclear physics; oscillations and waves; optics; electromagnetism; electric circuits; thermodynamics. Throughout, exercises are cross-referenced to emphasise the relationships between the different topics covered. There is also a section listing physical constants and other useful data, including a summary of some important mathematical results.

In discussing the key factors and most suitable methods of approach for given problems, this book imparts many useful insights, and will be invaluable to anyone taking first- or second-year undergraduate courses in physics.

Physics by Example

Physics by Example

200 Problems and Solutions

W. G. REES
Scott Polar Research Institute, University of Cambridge

CAMBRIDGE
UNIVERSITY PRESS

Published by the Press Syndicate of the University of Cambridge
The Pitt Building, Trumpington Street, Cambridge CB2 1RP
40 West 20th Street, New York, NY 10011-4211, USA
10 Stamford Road, Oakleigh, Melbourne 3166, Australia

First published 1994

Printed in Great Britain at the University Press, Cambridge

A catalogue record for this book is available from the British Library

Library of Congress cataloguing in publication data

Rees, W. G.
Physics by example: 200 problems and solutions / W. G. Rees.
    p.   cm.
Includes index.
ISBN 0 521 44514 0. – ISBN 0 521 44975 8 (pbk.)
1. Physics – Problems, exercises etc. I. Title.
QC32.R383   1994
530′.076–dc20   93-34300 CIP

ISBN 0 521 44514 0 hardback
ISBN 0 521 44975 8 paperback

KT

For Christine

Contents

Preface

This is a book about physics, consisting of two hundred problems with solutions worked out in detail. My experience in teaching physics at university level has led me to believe that many first-year undergraduates find physics interesting but hard. With perhaps one or two exceptions the concepts are straightforward enough to be grasped, and students often enjoy learning these concepts, many of which are beautiful or at least intellectually satisfying. However, students often find that translating their understanding of the ideas of physics into problem-solving can be difficult. I believe the best way to learn how to solve problems is by example, hence this book. The level of difficulty of the problems (in some cases it would be more accurate to say the level of sophistication of the solutions) is intended to be roughly that of the first year of a physics course at a British university, and the range of topics treated is correspondingly intended to reflect typical first-year syllabuses. In order to ensure that this is so, I have drawn most of these problems from recent first-year examination papers at several British universities. However, I hope that it will find some use both below and above this level. Its intention is to help students to make the transition from school physics to university physics, so it assumes a background in physics and mathematics appropriate to such a level.

This is not a textbook in the normal sense of the word. Although I hope you will learn some physics from it, a book of this kind will inevitably have a number of defects. The first is that it is far too short to contain all the physics you are likely to meet even in a single year at university. I have deliberately kept it short to encourage you to work your way through all of it. However, even a much larger book of this kind, containing many more problems, would not be an adequate substitute for a textbook or a course of lectures. Lectures and textbooks communicate ideas, but this book is intended to develop skill in manipulating those ideas. I hope it will thus supplement, rather than duplicate, the function of a textbook. For these reasons I have largely avoided including problems that require little more than duplicating standard derivations, except where the results are of direct relevance to other problems or where I believe them to illustrate important ideas.

The arrangement of the text is roughly by subject area – dimensions, errors and statistical analysis; classical mechanics and dynamics; gravitation and orbits; special relativity; quantum, atomic and nuclear physics; oscillations and waves; optics; electromagnetism; electric circuits; thermodynamics – but some spillage from one area to another is inevitable. Perhaps it is also desirable, in that it might encourage students to realise that in 'real life' problems cannot always be neatly categorised. Within each section I have tried to group problems by topics, and also to try to put easier ones before more difficult ones and to try to introduce results in earlier problems for use in later problems. It has not always been possible to meet all these criteria simultaneously, and I have therefore cross-referenced the problems, and also provided an index. Generally speaking, the index does not contain entries for items whose location should be obvious from the structure of the book, nor does it contain entries for familiar and widely applicable topics such as the conservation of energy. There is also a section listing physical constants and other useful data, which also contains a compendium of mathematical results. This is not intended to be comprehensive, but to serve as a ready reference for the techniques necessary in solving the problems. Supplementary material, not strictly required in the solution of the problems, is enclosed by brackets [ ]. Marginal arrowheads ▶ show the positions of the answers.

The coverage of the book is not intended to be comprehensive. It does not purport to provide model solutions to every problem the student will meet, nor to cover every topic. However, I would be grateful to learn of any glaring omissions that readers may feel I have made.

Although I have tried to eliminate errors from the book, since I know they will be embarrassing, I expect that a few will remain. I should be very grateful if readers would inform me of any errors that they may discover.

W.G.R.
Cambridge

Acknowledgements

I wish to express my gratitude to a number of people who have contributed in various ways to this book. Caroline Roberts and Ian Simm encouraged me to write it. Daniel Sheard and John Liddicoat checked my solutions, though of course àny errors which remain are entirely my responsibility – anyone who has attempted to solve a problem for which someone else's solution was available will know that it is difficult not to agree with that solution. I would also like to acknowledge the unwitting contributions of the many undergraduates whose demands for clarity and understanding have kept me on my toes. Finally, I thank Christine Rees for her patience with me and the book.

Most of the problems in this book are not original, and I am grateful to the Universities of Birmingham, Cambridge, Nottingham and Oxford for granting me permission to use their material. Specifically, I acknowledge the University of Birmingham for problems 2, 11, 13, 14, 16, 19, 23, 24, 34, 39, 46, 50, 54, 58, 73, 82, 87, 101, 102, 112, 114, 125, 128, 139, 143, 152, 153, 154, 175, 187 and 190, the University of Nottingham for problems 6, 8, 9, 20, 25, 27, 30, 35, 42, 53, 66, 72, 78, 79, 81, 84, 85, 88, 89, 90, 94, 95, 97, 100, 103, 106, 116, 118, 120, 126, 127, 131, 136, 142, 144, 145, 147, 172, 173, 174, 194, 195, 197, 198 and 200, the University of Oxford for problems 21, 40, 69, 77, 83, 115, 117, 129, 140, 150, 157, 163, 164, 165, 169, 171, 177, 178, 192, 196 and 199, and the University of Cambridge for the remainder, excluding problems 3, 15, 37, 38, 41, 109, 110 and 111 for which I accept responsibility. Except in those few cases where problems have been devised especially for this book, the copyright in these problems remains the property of the universities supplying the problems. In many cases I have modified the problems, and I accept full responsibility for any inconsistencies or ambiguities which may have resulted from such modification.

Physical constants and other data

Speed of light *in vacuo*	c	$2.998 \times 10^8 \, \mathrm{m\,s^{-1}}$
Permeability of vacuum	ε_0	$8.854 \times 10^{-12} \, \mathrm{F\,m^{-1}}$
Permittivity of vacuum	μ_0	$4\pi \times 10^{-7} \, \mathrm{H\,m^{-1}}$
Planck constant	h	$6.626 \times 10^{-34} \, \mathrm{J\,s}$
Avogadro constant	N_A	$6.022 \times 10^{23} \, \mathrm{mol^{-1}}$
Molar gas constant	R	$8.314 \, \mathrm{J\,K^{-1}\,mol^{-1}}$
Boltzmann constant	k	$1.381 \times 10^{-23} \, \mathrm{J\,K^{-1}}$
Stefan–Boltzmann constant	σ	$5.671 \times 10^{-8} \, \mathrm{W\,m^{-2}\,K^{-4}}$
Gravitational constant	G	$6.673 \times 10^{-11} \, \mathrm{N\,m^2\,kg^{-2}}$
Charge on proton	e	$1.602 \times 10^{-19} \, \mathrm{C}$
Charge-to-mass ratio of electron	e/m_e	$1.759 \times 10^{11} \, \mathrm{C\,kg^{-1}}$
Mass of electron	m_e	$9.109 \times 10^{-31} \, \mathrm{kg}$ (511.0 keV)
Mass of proton	m_p	$1.673 \times 10^{-27} \, \mathrm{kg}$ (938.3 MeV)
Mass of neutron	m_n	$1.675 \times 10^{-27} \, \mathrm{kg}$ (939.6 MeV)
Unified atomic mass constant	m_u	$1.661 \times 10^{-27} \, \mathrm{kg}$
Compton wavelength of electron	λ_c	$2.426 \times 10^{-12} \, \mathrm{m}$
Acceleration due to gravity	g	$9.807 \, \mathrm{m\,s^{-2}}$
Standard atmospheric pressure		$1.013 \times 10^5 \, \mathrm{Pa}$
Earth's equatorial radius		$6.378 \times 10^6 \, \mathrm{m}$
Earth's mass		$5.978 \times 10^{24} \, \mathrm{kg}$
Mean Earth–Sun distance		$1.496 \times 10^{11} \, \mathrm{m}$
Sun's mass		$1.989 \times 10^{30} \, \mathrm{kg}$
1 light year		$9.461 \times 10^{15} \, \mathrm{m}$
Mean solar day		$8.640 \times 10^4 \, \mathrm{s}$
Sidereal day		$8.616 \times 10^4 \, \mathrm{s}$
Mean calendar year		$3.156 \times 10^7 \, \mathrm{s}$

SI prefixes

atto	a	10^{-18}
femto	f	10^{-15}
pico	p	10^{-12}
nano	n	10^{-9}
micro	μ	10^{-6}
milli	m	10^{-3}

kilo	k	10^3
mega	M	10^6
giga	G	10^9
tera	T	10^{12}

Mathematical notes

1 Solution of quadratic equations

$Ax^2 + Bx + C = 0$ has two solutions, which either are both real or are complex conjugates:

$$x = \frac{-B \pm \sqrt{(B^2 - 4AC)}}{2A}.$$

2 Numerical solution of equations

There are many numerical techniques for finding the root of $f(x) = 0$, i.e. the value of x for which the equation is true. One of the most widely applied is the *Newton–Raphson* (or *Newton*) method, according to which, if x_n is an approximation to the root, x_{n+1} is usually a better approximation, where

$$x_{n+1} = x_n - \frac{f(x_n)}{f'(x_n)}$$

and $f'(x) = df/dx$.

3 Trigonometric formulae

$$\sin(x + y) = \sin x \cos y + \cos x \sin y$$
$$\sin(x - y) = \sin x \cos y - \cos x \sin y$$
$$\cos(x + y) = \cos x \cos y - \sin x \sin y$$
$$\cos(x - y) = \cos x \cos y + \sin x \sin y$$
$$\sin(2x) = 2 \sin x \cos x$$
$$\cos(2x) = \cos^2 x - \sin^2 x = 2\cos^2 x - 1 = 1 - 2\sin^2 x$$
$$\cos^2 x + \sin^2 x = 1$$

$$\sin x = \frac{1}{2i}(\exp[ix] - \exp[-ix]) \quad \text{where } i^2 = -1$$

$$\cos x = \frac{1}{2}(\exp[ix] + \exp[-ix]) \quad \text{where } i^2 = -1$$

4 Hyperbolic functions

$$\sinh x = \frac{1}{2}(\exp x - \exp[-x])$$

$$\cosh x = \frac{1}{2}(\exp x + \exp[-x])$$

$$\tanh x = \frac{\sinh x}{\cosh x}$$

$$\cosh^2 x - \sinh^2 x = 1$$
$$\sinh(2x) = 2\sinh x \cosh x$$
$$\cosh(2x) = \cosh^2 x + \sinh^2 x = 2\cosh^2 x - 1 = 2\sinh^2 x + 1$$

In general, relationships between hyperbolic functions can be derived from the relationship between the corresponding trigonometric functions using

$$\sin(ix) = i\sinh x,$$
$$\cos(ix) = \cosh x.$$

5 Power series

The Binomial expansion often provides a simplification of more complicated expressions:

$$(1 + x)^n = 1 + nx + \frac{n(n - 1)}{2!}x^2 + \frac{n(n - 1)(n - 2)}{3!}x^3 + \dots .$$

If n is a positive integer the series contains $n + 1$ terms (up to x^n) and is valid for all values of x. If n is not a positive integer, the series is infinite, and converges if $|x| < 1$. Examples:

$$(1 + x)^4 = 1 + 4x + 6x^2 + 4x^3 + x^4,$$

$$(1 + x)^{-1/2} = 1 - \frac{x}{2} + \frac{3x^2}{8} - \frac{5x^3}{16} + \dots .$$

The following power series are also useful:

$$\exp x = 1 + x + \frac{x^2}{2!} + \frac{x^3}{3!} + \dots \qquad \text{(valid for all } x\text{)};$$

$$\ln(1 + x) = x - \frac{x^2}{2} + \frac{x^3}{3} - \frac{x^4}{4} + \dots \text{ (valid for } -1 < x \leqslant 1\text{)};$$

$$\cos x = 1 - \frac{x^2}{2!} + \frac{x^4}{4!} - \frac{x^6}{6!} + \dots \qquad \text{(valid for all } x\text{)};$$

$$\sin x = x - \frac{x^3}{3!} + \frac{x^5}{5!} - \frac{x^7}{7!} + \dots \qquad \text{(valid for all } x\text{)};$$

$$\tan x = x + \frac{x^3}{3} + \frac{2x^5}{15} + \dots \qquad \text{(valid for} -\pi/2 < x < \pi/2\text{)};$$

$$\cosh x = 1 + \frac{x^2}{2!} + \frac{x^4}{4!} + \frac{x^6}{6!} + \dots \qquad \text{(valid for all } x\text{)};$$

$$\sinh x = x + \frac{x^3}{3!} + \frac{x^5}{5!} + \frac{x^7}{7!} + \dots \qquad \text{(valid for all } x\text{)};$$

$$\tanh x = x - \frac{x^3}{3} + \frac{2x^5}{15} - \frac{17x^7}{215} + \dots \text{ (valid for all } x\text{)}.$$

The *Taylor series* describes the behaviour of the function $f(x)$ near x_0 as a power series in Δx, where $\Delta x = x - x_0$:

$$f(x) = f(x_0) + \frac{df}{dx} \Delta x + \frac{d^2 f}{dx^2} \frac{(\Delta x)^2}{2!} + \frac{d^3 f}{dx^3} \frac{(\Delta x)^3}{3!} + \dots .$$

The differential coefficients are all evaluated at $x = x_0$. The *Maclaurin series* is a Taylor series with $x_0 = 0$.

6 Random errors and statistics

6.1 Single variable

Given n independent measurements $x_1, x_2, \dots, x_n$ of some quantity, the *sample mean* is defined as

$$\langle x \rangle = \frac{1}{n} \sum_{i=1}^{n} x_i,$$

and the *sample variance* is defined as

$$s^2 = \frac{1}{n} \sum_{i=1}^{n} (x_i - \langle x \rangle)^2 = \frac{1}{n} \sum_{i=1}^{n} x_i^2 - \langle x \rangle^2,$$

s is the *standard deviation of the sample*. The best estimate of the *standard deviation in the mean* is

$$\sigma_{\mathrm{m}} \approx \frac{s}{\sqrt{(n-1)}},$$

and the result of n measurements is quoted as $\langle x \rangle \pm \sigma_{\mathrm{m}}$.

Given n measurements of a quantity x, $x_1 \pm \sigma_1$, $x_2 \pm \sigma_2$, $\ldots$, $x_n \pm \sigma_n$, the results can be combined to give a mean value

$$\langle x \rangle = \frac{\displaystyle\sum_{i=1}^{n} \frac{x_i}{\sigma_i^2}}{\displaystyle\sum_{i=1}^{n} \frac{1}{\sigma_i^2}}.$$

The standard deviation in the mean value is

$$\sigma_{\mathrm{m}} = \left(\sum_{i=1}^{n} \frac{1}{\sigma_i^2} \right)^{-1/2}.$$

If f is a function of a, b, c, $\ldots$, each of which has an associated uncertainty σ_a, σ_b, σ_c, $\ldots$, the uncertainty in f is given by

$$\sigma_f^2 = \left(\frac{\partial f}{\partial a} \right)^2 \sigma_a^2 + \left(\frac{\partial f}{\partial b} \right)^2 \sigma_b^2 + \left(\frac{\partial f}{\partial c} \right)^2 \sigma_c^2 + \ldots .$$

6.2 Probability distributions

If an event can occur with probability p at a single trial, the probability $P(n)$ of n such events in N independent trials is given by the *binomial distribution*:

$$P(n) = \frac{N!}{(N - n)!n!} p^n (1 - p)^{N-n}.$$

The expectation value of n is Np and the standard deviation is $[Np(1 - p)]^{1/2}$.

The limit of the binomial distribution as N tends to infinity and p tends to zero such that the expectation value of n is constant at μ is the *Poisson distribution*:

$$P(n) = \frac{\mu^n}{n!} \exp(-\mu).$$

The standard deviation of n is equal to the expectation value μ.

The limit of the binomial distribution as N tends to infinity is the *normal distribution*, which is expressed as a probability distribution function $p(x)\,dx$ such that the probability of x lying between x and $x + dx$ is $p(x)\,dx$:

$$p(x) = \frac{1}{\sigma\sqrt{(2\pi)}} \exp\left(- \frac{[x - \mu]^2}{2\sigma^2} \right).$$

Related to the normal distribution is the *error function*, defined as

$$\text{erf}(x) = \frac{2}{\sqrt{\pi}} \int_0^x \exp(-z^2) \, dz,$$

so that $\text{erf}(0) = 0$ and $\text{erf}(\infty) = 1$. The probability of observing a departure (positive or negative) of at least k standard deviations away from the mean is

$$1 - \text{erf} \frac{k}{\sqrt{2}}.$$

Table 1 gives some values of this function:

Table 1

k	$1 - \text{erf}(k/\sqrt{2})$	k	$1 - \text{erf}(k/\sqrt{2})$
0	1	2.5	0.0124
0.5	0.617	3.0	0.00270
1.0	0.317	3.5	4.7×10^{-4}
1.5	0.134	4.0	6.3×10^{-5}
2.0	0.0455	4.5	6.8×10^{-6}

Thus the probability that a measurement taken from a normal distribution will be within ± 2 standard deviations of the mean value is 95%, and it is conventional to regard deviations larger than this as significant.

6.3 *Linear regression*

If we have n pairs of measurements (x_i, y_i) and we wish to use them to find a linear relationship of the form

$$y = mx + c,$$

the standard procedure is to choose m and c such that the sum of the squares of the residuals

$$d_i = y_i - mx_i - c$$

is minimum. This is achieved by taking

$$m = \frac{\sum\limits_{i=1}^{n}(x_i - \langle x \rangle)y_i}{\sum\limits_{i=1}^{n}(x_i - \langle x \rangle)^2}$$

and

$$c = \langle y \rangle - m \langle x \rangle,$$

where $\langle x \rangle$ is the mean value of the x values, defined as

$$\langle x \rangle = \frac{1}{n} \sum_{i=1}^{n} x_i,$$

and similarly

$$\langle y \rangle = \frac{1}{n} \sum_{i=1}^{n} y_i.$$

It is assumed that all the error in the data is in the values of y, and that all the data points have equal weight. The uncertainty in the slope, Δm, is given by

$$(\Delta m)^2 = \frac{\displaystyle\sum_{i=1}^{n} d_i^2}{D(n-2)},$$

and the uncertainty in the intercept, Δc, is given by

$$(\Delta c)^2 = \left(\frac{1}{n} + \frac{\langle x \rangle^2}{D} \right) \frac{\displaystyle\sum_{i=1}^{n} d_i^2}{n-2},$$

where

$$D = \sum_{i=1}^{n} (x_i - \langle x \rangle)^2.$$

7 Differentiation

$$\frac{d}{dx} x^n = n x^{n-1} \qquad \text{(for } n \text{ constant)}$$

$$\frac{d}{dx} \exp x = \exp x$$

$$\frac{d}{dx} \ln x = \frac{1}{x}$$

$$\frac{d}{dx} \sin x = \cos x$$

$$\frac{d}{dx} \cos x = -\sin x$$

$$\frac{d}{dx}\tan x = \frac{1}{\cos^2 x}$$

$$\frac{d}{dx}\sinh x = \cosh x$$

$$\frac{d}{dx}\cosh x = \sinh x$$

$$\frac{d}{dx}\tanh x = \frac{1}{\cosh^2 x}$$

$$\frac{d}{dx}[af(x)] = a\frac{df}{dx} \qquad \text{(for } a \text{ constant)}$$

$$\frac{d}{dx}f(ax) = af'(ax) \qquad \text{(for } a \text{ constant and } f' = df/dx)$$

$$\frac{d}{dx}[f(x)g(x)] = f(x)\frac{dg}{dx} + g(x)\frac{df}{dx}$$

$$\frac{d}{dx}\left[\frac{f(x)}{g(x)}\right] = \frac{1}{g}\frac{df}{dx} - \frac{f}{g^2}\frac{dg}{dx}$$

$$\frac{d}{dx}[f(g(x))] = \frac{df}{dz}\frac{dz}{dx} \qquad \text{(where } z = g(x))$$

8 Integration

The constants of integration have been omitted from the indefinite integrals.

$$\int x^n \, dx = \frac{x^{n+1}}{n+1} \qquad \text{(for } n \text{ constant; } n \neq -1)$$

$$\int \exp x \, dx = \exp x$$

$$\int \frac{dx}{x} = \ln x$$

$$\int \ln x \, dx = x \ln x - x$$

$$\int \sin x \, dx = -\cos x$$

$$\int \cos x \, dx = \sin x$$

$$\int \tan x \, dx = -\ln \cos x$$

$$\int_0^{\pi/2} \cos x \sin x \, dx = \frac{1}{2}$$

$$\int_0^{\pi/2} \cos^2 x \sin x \, dx = \frac{1}{3}$$

$$\int_0^\infty \exp(-ax^2) \, dx = \frac{1}{2}\sqrt{\frac{\pi}{a}} \qquad \text{(for } a \text{ constant)}$$

$$\int_0^\infty x \exp(-ax^2) \, dx = \frac{1}{2a} \qquad \text{(for } a \text{ constant)}$$

$$I_n = \int_0^\infty x^n \exp(-ax^2) \, dx = \frac{n-1}{2a} I_{n-2} \qquad \text{(for } n, a \text{ constant; } n \geqslant 2)$$

$$\int f(x)g(x) \, dx = f(x)\int g(x) \, dx - \int \left[\int g(x) \, dx\right]\left(\frac{df}{dx}\right) dx$$

$$\int f(ax) \, dx = \frac{1}{a}\int f(z) \, dz \qquad \text{(for } a \text{ constant; } z = ax)$$

9 Vector algebra

If

$$\mathbf{A} = (A_x, A_y, A_z)$$

and

$$\mathbf{B} = (B_x, B_y, B_z)$$

are vectors described by their Cartesian components, their *scalar product* is

$$\mathbf{A} \cdot \mathbf{B} = A_x B_x + A_y B_y + A_z B_z = |\mathbf{A}|\,|\mathbf{B}| \cos \theta,$$

where

$$|\mathbf{A}| = (A_x^2 + A_y^2 + A_z^2)^{1/2}$$

is the *modulus* of $\mathbf{A}$ (and similarly for $|\mathbf{B}|$), and θ is the angle between the two vectors. The *vector product* is

$$\mathbf{A} \times \mathbf{B} = (A_y B_z - A_z B_y, A_z B_x - A_x B_z, A_x B_y - A_y B_x).$$

The modulus of $\mathbf{A} \times \mathbf{B}$ is $|\mathbf{A}|\,|\mathbf{B}| \sin \theta$ and the direction of $\mathbf{A} \times \mathbf{B}$ is normal to the plane containing $\mathbf{A}$ and $\mathbf{B}$.

The differential operators *grad*, *div* and *curl* are defined as follows in Cartesian coordinates.

$$\text{grad } f = \nabla f = \left(\frac{\partial f}{\partial x}, \frac{\partial f}{\partial y}, \frac{\partial f}{\partial z}\right),$$

where f is a scalar. Thus grad operates on a scalar to produce a vector.

$$\text{div } \mathbf{A} = \mathbf{\nabla} \cdot \mathbf{A} = \frac{\partial A_x}{\partial x} + \frac{\partial A_y}{\partial y} + \frac{\partial A_z}{\partial z},$$

where $\mathbf{A}$ is the vector (A_x, A_y, A_z). Thus div operates on a vector to produce a scalar.

$$\text{curl } \mathbf{A} = \mathbf{\nabla} \times \mathbf{A} = \left(\frac{\partial A_z}{\partial y} - \frac{\partial A_y}{\partial z}, \frac{\partial A_x}{\partial z} - \frac{\partial A_z}{\partial x}, \frac{\partial A_y}{\partial x} - \frac{\partial A_x}{\partial y} \right).$$

Thus curl operates on a vector to produce another vector.

10 Differential equations

10.1 Undamped simple harmonic motion

$$\frac{d^2x}{dt^2} + kx = 0,$$

where k is a positive constant, has the general solution

$$x = A \cos(t\sqrt{k}) + B \sin(t\sqrt{k}),$$

where A and B are constants.

10.2 Damped simple harmonic motion

$$\frac{d^2x}{dt^2} + c\frac{dx}{dt} + kx = 0,$$

where c and k are constants. The type of solution depends on the value of $D = k - c^2/4$.

(a) $D > 0$ (under-damped). The general solution is

$$x = (A \cos[t\sqrt{D}] + B \sin[t\sqrt{D}]) \exp(-ct/2),$$

where A and B are constants.

(b) $D = 0$ (critically damped). The general solution is

$$x = (A + Bt) \exp(-ct/2),$$

where A and B are constants.

(c) $D < 0$ (over-damped). The general solution is

$$x = A \exp[(-c/2 + \sqrt{[-D]})t] + B \exp[(-c/2 - \sqrt{[-D]})t],$$

where A and B are constants.

10.3 Exponential growth or decay

$$\frac{dx}{dt} = ax + b,$$

where a and b are constants, has the general solution

$$x = A \exp(at) - \frac{b}{a},$$

where A is a constant.

Hints for solving physics problems

As I have written in the preface, I believe that the best way of learning how to solve physics (or any other) problems is by practice and from example. However, there are a few basic hints, all of which are really just common sense and therefore fairly obvious.

- Try to understand the physics of the problem before launching into a mathematical analysis. It will sometimes be possible to obtain a rough preliminary numerical answer, perhaps using simplifying assumptions.
- Lay your work out neatly on the page, without being obsessive about it, and explain what you are doing and why you are doing it.
- Draw a diagram if it helps (it nearly always does).
- Try to keep expressions algebraic rather than numerical. This may mean that you have to invent symbols for quantities that are specified numerically. But it has the advantage that the dimensions of your answer can be checked at the end of the calculation. It is also much less likely that you will make mistakes if you are manipulating a few symbols rather than strings of digits. Finally, if you can express your answer algebraically you may be able to check that the variables affect the answer in the way you expect.
- Check the dimensions of your answer if this is possible and appropriate.
- Check the magnitude of your answer against common sense or other knowledge, if possible.
- Calculate the error (uncertainty) in your answer, perhaps just as a rough estimate, if you have the information to do it.
- If your numerical answer to a problem does not agree with a value which you are reasonably sure is correct, and all other attempts at identifying the error have failed, look at the ratio of your answer to the 'correct' one. You may find that your answer is wrong by a simple factor such as 2, π, $\ln 2$ etc. If so, go back to your working and try to find where you introduced the error, but remember that the error could be in the 'correct' answer, and not in your solution. (Some teachers may dislike this suggestion, but I have found it useful as a last resort!)

A note about significant figures

There is a well established convention that, if no other indication of the uncertainty in an experimental value is given, the least significant figure is assumed to be correct to within ± 0.5 digit. Thus, for example, 1.47 is taken to mean 1.47 ± 0.005. This leads to a convenient rule of thumb that if a set of experimental data are specified to N significant figures, a result obtained by combining the data will also be valid to N significant figures. However, this simple rule must be used with caution since it is not as accurate as a rigorous error calculation. Some examples will make this clear.

(i) If data are subtracted, the number of significant figures can be reduced. For example, $10.9 - 3.07$ (both specified to three significant figures) can not properly be evaluated as 7.83 but only as 7.8.

(ii) The fractional error of a value specified to N significant figures depends not only on N but also on the value itself. For example, 1.03 is accurate to $\pm 0.005/1.03 = 0.5\%$ whereas 9.87 is accurate to $\pm 0.005/9.87 = 0.05\%$. This can lead to apparent inconsistencies if the rule of not specifying results to more significant figures than are justified by the data is followed rigorously. For example, if $X = 3.7$ (two significant figures), $X/2$ would be written as 1.9 to the same number of significant figures, obscuring the fact that the latter quantity is exactly half the former.

(iii) Intermediate results should be calculated to one or two more significant figures than are justified by the data, otherwise rounding errors can build up. For example, the reciprocal of 9.57, calculated to the same number of significant figures (three), is 0.104, but the reciprocal of 0.104 calculated to three significant figures is 9.62. However, if we had written $1/9.57 = 0.1045$ and then taken the reciprocal to three significant figures, we would have retrieved the original value of 9.57.

Problems and solutions

Dimensions, errors and statistical analysis

Problem 1

The viscosity η of a gas depends on the mass, the effective diameter and the mean speed of the molecules. Use dimensional analysis to find η as a function of these variables. Hence estimate the diameter of the methane (CH_4) molecule given that η has values of $2.0 \times 10^{-5}\,\mathrm{kg\,m^{-1}\,s^{-1}}$ for helium and $1.1 \times 10^{-5}\,\mathrm{kg\,m^{-1}\,s^{-1}}$ for methane at room temperature, and that the diameter of the helium atom is 2.1×10^{-10} m. [Relative atomic masses for C and He are 12 and 4 respectively.]

Solution

Assume that

$$\eta = km^\alpha d^\beta v^\gamma,$$

where k, α, β and γ are dimensionless constants, m is the mass, d the diameter and v the mean speed of a molecule. The dimensions of the mass m are M, those of the diameter d are L and those of speed v are $\mathrm{LT^{-1}}$. From the units given in the problem (or from our own knowledge) the dimensions of viscosity are $\mathrm{ML^{-1}T^{-1}}$. Thus

$$\mathrm{ML^{-1}T^{-1}} = \mathrm{M^\alpha L^\beta L^\gamma T^{-\gamma}},$$

so, by inspection,

$$\alpha = 1,$$
$$\beta = -2,$$
$$\gamma = 1,$$

and hence

▶ $$\eta = k\frac{mv}{d^2}.$$

Now for a gas we expect mv^2 to be proportional to the absolute temperature T (the root mean square speed c is given by $mc^2 = 3k_B T$, where k_B is Boltzmann's constant, and v is proportional to c), so at a fixed temperature T the velocity v should be proportional to $m^{-1/2}$.

Hence we expect

$$\eta d^2 m^{-1/2}$$

to be constant at fixed temperature, and thus

$$d_{\mathrm{CH_4}} = d_{\mathrm{He}}\left[\frac{\eta_{\mathrm{He}}}{\eta_{\mathrm{CH_4}}}\right]^{1/2}\left[\frac{m_{\mathrm{CH_4}}}{m_{\mathrm{He}}}\right]^{1/4}$$

$$= 2.1 \times 10^{-10}(2.0/1.1)^{1/2}(16/4)^{1/4} \text{ metres}$$

$$= 4.0 \times 10^{-10} \text{ m}.$$

This seems a reasonable answer. We expect the methane molecule to be bigger than the helium atom, although not by much.

Problem 2

Use the method of dimensions to obtain the form of the dependence of the lift force per unit wingspan on an aircraft wing of width (in the direction of motion) L, moving with velocity v through air of density ρ, on the parameters L, v and ρ.

Solution

Figure 1 shows the width of the wing and the wingspan.

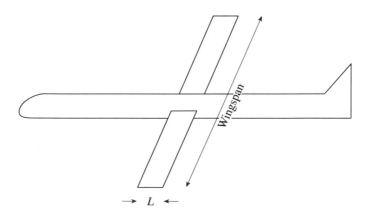

Figure 1

Let us call the lift per unit wingspan Φ, and write

$$\Phi = kL^\alpha v^\beta \rho^\gamma,$$

where k, α, β and γ are dimensionless constants. Since the dimensions of force are MLT^{-2}, the dimensions of Φ are MT^{-2}. Thus

$$MT^{-2} = L^\alpha L^\beta T^{-\beta} M^\gamma L^{-3\gamma}.$$

So by equating the terms in M, $\gamma = 1$.
By equating the terms in T, $-\beta = -2$ so $\beta = 2$.
By equating the terms in L, $\alpha + \beta - 3\gamma = 0$, therefore $\alpha = 1$.
    Thus we may write

▶ $$\Phi = kLv^2\rho.$$

[This analysis suggests that we can define a dimensionless *lift coefficient* as

$$\frac{F}{\frac{1}{2}\rho v^2 Lw},$$

where w is the wingspan. The factor of $1/2$ is conventional. As an example, a Boeing 707 has a maximum mass of about 148×10^3 kg and a wing area of about 280 m^2. At $35\,000$ feet altitude ($10\,700$ m), where the air density ρ is about 0.37 kg m^{-3}, the cruising speed is about 250 m s^{-1}. The lift coefficient under these conditions is thus about

$$\frac{2 \times 148 \times 10^3 \times 9.8}{0.37 \times 250^2 \times 280} = 0.45.$$

Most aircraft wings have lift coefficients in the range 0.2 to 0.6, depending on the design of the wing and its orientation with respect to the aircraft's motion.]

Problem 3

The drag force per unit length, F, acting on a cylinder of diameter D moving at velocity v perpendicularly to its axis through a fluid of density ρ and viscosity η depends only on D, ρ, η and v. Use the data in Table 2 to estimate the force per unit length acting on a cylinder of diameter 0.1 m moving at 61 m s^{-1} through air.

Table 2

fluid	density ρ kg m^{-3}	viscosity η N s m^{-2}	cylinder diameter m	velocity v m s^{-1}	drag force per unit length F N m^{-1}
water	998	1.00×10^{-3}	0.01	30	4500
mercury	13 546	1.55×10^{-3}	0.01	5.5	700
air	1.30	1.83×10^{-5}	0.10	61	?

Solution

If we write down the dimensions of the various quantities involved,

$$[F] = MT^{-2},$$
$$[D] = L,$$
$$[\rho] = ML^{-3},$$
$$[\eta] = ML^{-1}T^{-1},$$
$$[v] = LT^{-1},$$

we can see that we will be unable to find a simple expression for F in terms of D, v, ρ and η, since we will obtain three equations (one each from mass, length and time) for four unknowns. We thus need to look for two *dimensionless groups*.

It is clear from these dimensions that

$$A = \frac{F}{\rho v^2 D}$$

has no dimensions, nor has

$$B = \frac{\rho v D}{\eta}.$$

We assume that there is a unique relationship (which may not be a simple one) between the dimensionless variables A and B.

For the cylinder in water, combination A has a value of 0.501 and combination B is 2.99×10^5.

For the cylinder in mercury, A is 0.171 and B is 4.81×10^5.

For the cylinder in air, B has a value of 4.33×10^5 which is about 74% of the way from the value for water to the value for mercury. We might reasonably assume that A will have a value close to $0.26 \times 0.501 + 0.74 \times 0.171 \approx 0.26$ (see figure 2). [We know that there is a unique relationship between A and B, but we are only guessing that the relationship is approximately linear in the range $0.171 \leqslant A \leqslant 0.501$.]

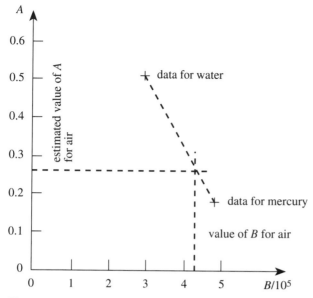

Figure 2

Thus we expect that F will be approximately

$$0.26 \times 1.30 \times (61)^2 \times 0.10 \, \mathrm{N\,m^{-1}}$$
$$= 1.3 \times 10^2 \, \mathrm{N\,m^{-1}}.$$

[The constant A is in fact half the drag coefficient, and B is the Reynolds number. We do not need to know this to solve the problem, and the choice of dimensionless groups is not unique. For example, $1/A$ and $1/B$ would have been equally valid, as would AB and A/B. However, we do know that there can only be two independent groups of variables because the total number of variables (5) exceeds the total number of dimensions (3) by two.]

Problem 4

A recent high-precision determination of g has a quoted error of 6 parts in 10^9. Estimate the increase in height at the Earth's surface which gives a change in g equal to this error. If the dependence of g on geographical latitude at sea level is given by

$$g = g_0(1 + \beta \sin^2 \phi),$$

where ϕ is the latitude and β is a dimensionless constant with a value of 0.0053, estimate the northward displacement near latitude 45° which gives a change in g equal to the quoted error.

Solution

We need to know how the acceleration due to gravity, g, varies with height above the Earth's surface. Taking the Earth to be a spherically symmetric mass M, we can write

$$g = \frac{GM}{r^2},$$

where r is the distance from the centre, assuming that r is greater than the Earth's radius. Differentiating both sides gives

$$dg = \frac{-2GM}{r^3}\,dr.$$

Dividing by g (and substituting from our previous expression for g) gives

$$\frac{dg}{g} = -\frac{2dr}{r}$$

(which, apart from the $-$ sign, we recognise as the familiar result from error analysis). Thus the fractional change in r is half the fractional change in g (ignoring the minus sign), so the change in r required to achieve a fractional change of 6×10^{-9} in g must be

$$\tfrac{1}{2} \times 6 \times 10^{-9} \times 6.4 \times 10^6 \text{ m}$$

(we have taken the Earth's radius as 6400 km)

$$= 0.019 \text{ m}.$$

▶   I.e. a change in height of 19 mm would cause a change in g of 6 parts in 10^9.

To consider the effect of a change of latitude, we differentiate the expression $g = g_0(1 + \beta \sin^2 \phi)$ with respect to the latitude ϕ to obtain $dg/d\phi = 2g_0\beta \sin \phi \cos \phi$. Thus

$$\Delta\phi = \frac{\Delta g}{2g_0\beta \sin \phi \cos \phi}.$$

Taking $\Delta g/g = 6 \times 10^{-9}$, $\beta = 0.0053$ and $\phi = \pi/4$ radians gives $\Delta\phi = 1.13 \times 10^{-6}$ radians. Multiplying this by the Earth's radius of

▶ 6.4×10^6 m gives a distance on the Earth's surface of about 7 m.

Problem 5

The experimental values in Table 3 were obtained for the Young modulus E and the shear modulus G of steel:

Table 3

$E/10^{10}\,\mathrm{N\,m^{-2}}$	21.1	21.0	20.9	20.6
$G/10^{10}\,\mathrm{N\,m^{-2}}$	8.12	8.15	8.13	8.08

The formula $(E/2G) - 1 = v$ was then used to calculate the Poisson ratio v. Find v and estimate its error.

Solution

If the errors in E and G are uncorrelated, we should be able to obtain equally valid results by calculating

$$v = \frac{\langle E \rangle}{2 \langle G \rangle} - 1$$

and

$$v = \left\langle \frac{E}{2G} - 1 \right\rangle,$$

where $\langle x \rangle$ is the mean value of x. As a check, we will try both methods.

Method 1

Working in units of $10^{10}\,\mathrm{N\,m^{-2}}$, calculate

$$\langle E \rangle = (21.1 + 21.0 + 20.9 + 20.6)/4$$
$$= 20.9$$

and

$$\langle E^2 \rangle = (21.1^2 + 21.0^2 + 20.9^2 + 20.6^2)/4$$
$$= 436.845.$$

So the square root of the sample variance is

$$(436.845 - 20.90^2)^{1/2}$$
$$= 0.19.$$

The error in the mean value $\langle E \rangle$ is thus

$$0.19/\sqrt{3} = 0.11,$$

so the fractional error in $\langle E \rangle$ is

$$0.11/20.9 = 0.5\%.$$

Similarly, $\langle G \rangle = 8.12$ and the fractional error in $\langle G \rangle$ is 0.2%. Combining these,

$$\langle E \rangle / 2 \langle G \rangle = 1.287 \pm 0.5\%$$
$$= 1.287 \pm 0.006,$$

so

▶ $$v = 0.287 \pm 0.006 \text{ (no units)}.$$

Method 2

Calculate $E/2G - 1$ for each set of values (Table 4).

Table 4

$E/10^{10}$ N m^{-2}	21.1	21.0	20.9	20.6
$G/10^{10}$ N m^{-2}	8.12	8.15	8.13	8.08
$E/2G - 1$	0.299	0.288	0.285	0.275

Evaluating the mean value of the quantity and of its square as before, and obtaining the error in the mean, gives

▶ $$v = 0.287 \pm 0.005.$$

This value is not significantly different from the result of method 1, so we are justified in assuming that the errors in E and G are uncorrelated.

Problem 6

The thickness t of an aluminium foil is to be determined using an experiment involving the absorption of β-particles. The sample of aluminium is placed between a β-source and a β-detector and the measured count rate varies according to the thickness of the aluminium foil as follows:

$$n = n_0 \exp(-\mu t).$$

Given that the number n_0 of counts recorded in an interval of time in the absence of the foil is 572 and the number n of counts in the presence of the foil is 417, calculate the thickness of the foil and the error in this quantity. μ is a constant with a value of $(1.38 \pm 0.05) \times 10^3 \text{ m}^{-1}$.

In a second experiment using a different technique the thickness, determined from five repeated measurements, is found to be 0.29 mm with a standard deviation of 0.1 mm. Comment upon whether the results of the two experiments are consistent.

For a material with a given value of μ, what thickness can be determined, using this radiation-counting technique, with the smallest percentage error?

Solution

Rearrange the equation to make t the subject:

$$t = \frac{1}{\mu} \ln \frac{n_0}{n}.$$

Now we know that in counting a large number N of discrete random events, the standard deviation is $\sqrt{N}$, so the fractional error in n_0 is

$$\frac{\sqrt{572}}{572} = 4.2\%.$$

Similarly the fractional error in n is

$$\frac{\sqrt{417}}{417} = 4.9\%.$$

Thus the fractional error in n_0/n is $(4.2^2 + 4.9^2)^{1/2}\% = 6.5\%$, so

$$\frac{n_0}{n} = \frac{572}{417} = 1.372 \pm 6.5\%.$$

Now if $y = \ln x$, $dy = dx/x$ so the *absolute* error in $\ln(n_0/n)$ is equal to the *fractional* error in n_0/n. Thus

$$\ln \frac{n_0}{n} = \ln 1.372 \pm 0.065$$
$$= 0.316 \pm 0.065$$
$$= 0.32 \pm 21\%.$$

The fractional error in μ is $0.05/1.38 = 3.6\%$, so the fractional error in t is

$$(21^2 + 3.6^2)^{1/2}\% = 21\%.$$

Thus

$$t = \frac{0.32}{1.38 \times 10^3}\,\text{m} = 0.23\,\text{mm} \pm 21\%$$

▶ $\qquad = 0.23 \pm 0.05\,\text{mm}.$

The second experiment determined t to be 0.29 ± 0.10 mm. To see whether these two results are consistent we calculate the difference between them, which is 0.06 mm, and the standard deviation of this difference, which is $(0.05^2 + 0.10^2)^{1/2}$ mm $= 0.11$ mm.

In order to reject the hypothesis that a variable has a value of zero, we usually require that the modulus of the ratio of the value to its standard deviation should be at least 2. In this case it is only $0.06/0.11 = 0.5$, so we cannot reject the hypothesis. The two measurements of thickness are thus consistent. [This is actually obvious without the calculation, since the range 0.29 ± 0.10 includes the value 0.23.]

[Since we have two independent measurements of the same quantity, we can combine them to obtain an improved estimate. The technique for doing this is to form a weighted average, the weights being the inverse squares of the errors. Thus

$$\langle t \rangle = \frac{0.23 \times 0.05^{-2} + 0.29 \times 0.10^{-2}}{0.05^{-2} + 0.10^{-2}}\,\text{mm}$$

$$= 0.24\,\text{mm}.$$

The error in $\langle t \rangle$ is the inverse square of the sum of the weights, i.e. $(0.05^{-2} + 0.10^{-2})^{-1/2} = 0.05$ mm, so the best value for t becomes
▶ 0.24 ± 0.05 mm.]

Returning to the radiation technique for measuring thickness, and writing Δn for the error in n, and similarly for other quantities, we may put

$$\frac{\Delta n_0}{n_0} = \frac{1}{\sqrt{n_0}}$$

and

$$\frac{\Delta n}{n} = \frac{1}{\sqrt{n}}$$

as before, so the fractional error in n_0/n, which is also the absolute error in $\ln(n_0/n)$, must be

$$(1/n_0 + 1/n)^{1/2}.$$

The fractional error in $\ln(n_0/n)$ is thus

$$\frac{(1/n_0 + 1/n)^{1/2}}{\ln\left(\dfrac{n_0}{n}\right)}.$$

Now since the value of μ is known, its fractional error is constant, so the fractional error in t will be smallest when the fractional error in $\ln(n_0/n)$ is smallest. If we call this F, and assume that n_0 is a constant (determined by the radiation source and the time for which the detector is operated), the smallest fractional error in t is found at the value of n for which $\partial F/\partial n$ is zero.

Differentiating our expression for F, and setting it equal to zero, gives

$$\ln\left(\frac{n_0}{n}\right) = 2\left(1 + \frac{n}{n_0}\right).$$

This must be solved graphically or numerically. We will try the Newton–Raphson method. Putting $x = (n_0/n)$, we need to find the root of

$$f(x) = \ln x - 2/x - 2 = 0.$$

In general, if x_n is an approximation to the root,

$$x_{n+1} = x_n - \frac{f(x_n)}{f'(x_n)}$$

is a better approximation. Since $f'(x) = 1/x + 2/x^2$, this can be written

$$x_{n+1} = x_n - \frac{\ln x_n - 2/x_n - 2}{1/x_n + 2/x_n^2} = \frac{3x_n^2 + 4x_n - x_n^2 \ln x_n}{2 + x_n}.$$

Let us take $x_0 = 1$ as our initial approximation. Applying the Newton–Raphson method gives

$$x_1 = 2.33,$$
$$x_2 = 4.86,$$
$$x_3 = 7.72,$$
$$x_4 = 9.04,$$
$$x_5 = 9.18,$$
$$x_6 = 9.19,$$
$$x_7 = 9.19.$$

The solution is thus

$$(n_0/n) \approx 9.19,$$

so the thickness t which can be determined with the smallest fractional error is

> $(\ln 9.19)/\mu$
> $= 2.22/\mu.$

[This has been rather a lengthy calculation, so perhaps we should check that the answer is correct. We were looking for the solution of

$$f(x) = \ln x - 2/x - 2 = 0,$$

and have obtained a value of $x = 9.19$. Let us check by evaluating $f(9.19)$:

$$f(9.19) = 0.0005.$$

This is close to zero, but to be sure that our answer is to be correct to three significant figures we should check that neither $f(9.18)$ nor $f(9.20)$ is closer to zero. Evaluating these, we find

$$f(9.18) = -0.0008,$$
$$f(9.20) = 0.0018.$$

Thus we can be sure that $x = 9.19$ is the solution, correct to three significant figures.]

Problem 7

The density of a uniform cylinder is determined by measuring its mass m, length l and diameter d. Calculate the density (in $kg\,m^{-3}$) and its error from the following values:

> $m = 47.36 \pm 0.01$ g;
> $l = 15.28 \pm 0.05$ mm;
> $d = 21.37 \pm 0.04$ mm.

Solution

Clearly the formula for the density ρ in terms of the measured quantities is

$$\rho = \frac{4m}{\pi l d^2}.$$

Straightforward substitution of the measured values gives $\rho = 8642 \text{ kg m}^{-3}$.

The fractional error in m is $0.01/47.36 = 0.02\%$.
The fractional error in l is $0.05/15.28 = 0.33\%$.
The fractional error in d^2 is $2(0.04/21.37) = 0.37\%$.

So the fractional error in ρ is $(0.02^2 + 0.33^2 + 0.37^2)^{1/2}\% = 0.50\%$.

▶ Thus $\rho = (8.64 \pm 0.04) \times 10^3 \text{ kg m}^{-3}$.

Problem 8

In an experiment to determine the charge-to-mass ratio e/m of the electron using a cylindrical diode, the following equation is derived:

$$\frac{e}{m} = \frac{8V}{r^2 B^2},$$

where V is the potential difference across the diode of radius r at the critical magnetic field B. The latter is supplied by passing a current I through a solenoid of diameter D and length L, and is given by

$$B = \frac{\mu_0 n I}{(1 + D^2/L^2)^{1/2}},$$

where n is the number of turns per unit length (see problem 138). The appropriate experimental quantities are determined as follows:

$n = 3920 \text{ m}^{-1}$;
$D = 0.035 \pm 0.001 \text{ m}$;
$L = 0.120 \pm 0.001 \text{ m}$;
$r = (3.4 \pm 0.1) \times 10^{-3} \text{ m}$;
$I = 1.92 \pm 0.02 \text{ A}$;
$V = 20 \pm 1 \text{ V}$.

(i) Determine the magnitude of the field B and the error in this quantity.
(ii) Determine the value of e/m and the error in this quantity.

Solution

(i) The fractional error in D^2 is $2(0.001/0.035) = 5.7\%$.
The fractional error in L^2 is $2(0.001/0.120) = 1.7\%$.
So the fractional error in D^2/L^2 is $(5.7^2 + 1.7^2)^{1/2}\% = 6.0\%$.

Calculating the value of D^2/L^2 gives 0.0851, so the absolute error in this quantity is 0.005. Thus

$$1 + D^2/L^2 = 1.085 \pm 0.005$$
$$= 1.085 \pm 0.5\%,$$

so

$$(1 + D^2/L^2)^{1/2} = 1.042 \pm 0.25\%.$$

The fractional error in I is 1.0%, and we may assume that there is no error in the values of μ_0 or of n, so the fractional error in B must be $(0.25^2 + 1.0^2)^{1/2} = 1.0\%$. Thus (taking $\mu_0 = 4\pi \times 10^{-7}\,\mathrm{H\,m^{-1}}$) we obtain

$$B = 9.08\,\mathrm{mT} \pm 1.0\%$$
$$= 9.08 \pm 0.09\,\mathrm{mT}.$$

(ii) Fractional error in V is $1/20 = 5.0\%$.
Fractional error in r^2 is $2(0.1/3.4) = 5.9\%$.
Fractional error in B^2 is $2 \times 1.0\% = 2.0\%$.
So fractional error in e/m is $(5.0^2 + 5.9^2 + 2.0^2)^{1/2}\% = 8.0\%$.
Thus

$$e/m = 1.68 \times 10^{11}\,\mathrm{C\,kg^{-1}} \pm 8.0\%$$
$$= (1.7 \pm 0.1) \times 10^{11}\,\mathrm{C\,kg^{-1}}.$$

We can compare this with the currently accepted value for the charge-to-mass ratio of the electron, which is $1.759 \times 10^{11}\,\mathrm{C\,kg^{-1}}$. It is clear that the experimental value is consistent with this value to within the experimental error.

Problem 9

A pendulum consists of a copper sphere of radius R and density ρ suspended from a string. The motion of the sphere experiences a viscous drag from the air such that the amplitude of oscillation, A, decays with time, t, as follows:

$$A = A_0 \exp(-\gamma t),$$

where

$$\gamma = \frac{9\eta}{4R^2\rho},$$

A_0 is the amplitude at time $t = 0$ and η is the viscosity of the air. The measurement of the amplitudes is accurate to 1%; other measurements are recorded below. Evaluate the time taken for the amplitude to fall to 85% of A_0 and the error in this quantity. Which experimental parameter contributes the largest error to the final result?

$$\eta = (1.78 \pm 0.02) \times 10^{-5} \, \text{kg m}^{-1}\text{s}^{-1}.$$
$$R = 5.2 \pm 0.2 \, \text{mm}.$$
$$\rho = (8.92 \pm 0.05) \times 10^3 \, \text{kg m}^{-3}.$$

Solution

We want to determine t, so we rearrange the equations to make t the subject:

$$t = \frac{1}{\gamma} \ln \frac{A_0}{A} = \frac{4R^2\rho}{9\eta} \ln \frac{A_0}{A}.$$

A and A_0 are each determined to a fractional accuracy of 1.0%, so the ratio A_0/A is determined to a fractional accuracy of $(1^2 + 1^2)^{1/2}\%$ $= 1.4\%$.

If $y = \ln x$, $dy = dx/x$, so the absolute accuracy in $\ln x$ is equal to the fractional accuracy in x. Thus $\ln(A_0/A)$ is subject to an absolute accuracy of ± 0.014, and since $A_0/A = 100/85$, $\ln(A_0/A) = 0.163 \pm 0.014$, i.e. a fractional accuracy of 8.6%.

The fractional error in R^2 is $2(0.2/5.2) = 7.7\%$.
The fractional error in ρ is $0.05/8.92 = 0.6\%$.
The fractional error in η is $0.02/1.78 = 1.1\%$.

Thus the fractional error in t is $(8.6^2 + 7.7^2 + 0.6^2 + 1.1^2)^{1/2}\% = 11.6\%$.
The value of t is

$$\frac{4(5.2 \times 10^{-3})^2(8.92 \times 10^3)}{9(1.78 \times 10^{-5})} \ln \frac{100}{85} \, \text{s}$$
$$= 979 \, \text{s}.$$

So

$$t = 979 \, \text{s} \pm 11.6\%$$
$$= (1.0 \pm 0.1) \times 10^3 \, \text{s}.$$

The largest contribution to the error in t clearly comes from the error in determining the amplitudes A_0 and A.

Problem 10

A flat circular ring has mass M, outer radius a and inner radius b (see figure 3).

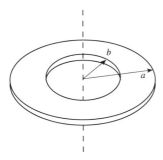

Figure 3

The measured values of these quantities are

$M = 0.191 \pm 0.003$ kg,
$a = 110 \pm 1$ mm,
$b = 15 \pm 1$ mm.

Find the moment of inertia of the ring about an axis through the centre, and normal to the plane of the ring, and estimate its error.

Solution

First we need to find the formula for I, the moment of inertia, in terms of the given quantities. If we consider a disc whose mass per unit area is σ and whose radius is r, its mass is $\pi\sigma r^2$ and its moment of inertia about an axis through the centre and normal to the plane of the disc is $\pi\sigma r^4/2$ (see problem 23). We can thus write the moment of inertia of the ring as

$$I = \frac{\pi\sigma}{2}(a^4 - b^4).$$

But we can also write the mass per unit area as

$$\sigma = \frac{M}{\pi(a^2 - b^2)},$$

so our expression for I becomes

$$I = \frac{M}{2}(a^2 + b^2).$$

Inserting the values given, we obtain

$$I = 0.191(0.110^2 + 0.015^2)/2 = 1.177 \times 10^{-3}\,\text{kg}\,\text{m}^2.$$

Now if the absolute error in a is Δa, its fractional error is $\Delta a/a$ and the fractional error in a^2 is $2\Delta a/a$. Thus the absolute error in a^2 is $2a\Delta a$, and similarly the absolute error in b is $2b\Delta b$, so the absolute error in $a^2 + b^2$ must be $(4a^2[\Delta a]^2 + 4b^2[\Delta b]^2)^{1/2}$. Evaluating this gives an absolute error of $2.2 \times 10^{-4}\,\text{m}^2$, and since $a^2 + b^2 = 1.23 \times 10^{-2}\,\text{m}^2$ this corresponds to a fractional error of 1.8%.

The fractional error in M is $(0.003/0.191) = 1.6\%$, so the fractional error in I is $(1.8^2 + 1.6^2)^{1/2}\% = 2.4\%$, thus

$$I = 1.177 \times 10^{-3}\,\text{kg}\,\text{m}^2 \pm 2.4\%$$
$$= (1.18 \pm 0.03) \times 10^{-3}\,\text{kg}\,\text{m}^2.$$

Problem 11

Write down in terms of the absolute errors Δl and Δb in l and b respectively expressions for the absolute errors in the following quantities:

    (i) l^2;
    (ii) b^2;
    (iii) $l^2 + b^2$;
    (vi) $\sqrt{(l^2 + b^2)}$;
    (v) $([l^2 + b^2]/12)^{1/2}$.

The moment of inertia of a bar, length l and breadth b, about an axis through the centre of mass (see figure 4), is mk^2, where

$$k = \sqrt{\frac{l^2 + b^2}{12}}$$

and m is the mass of the bar.

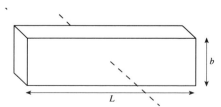

Figure 4

A particular bar has a length of about 1 m and a breadth of about 25 mm. It is required to find k to a precision of 1 part in 10^5. How accurately would you need to measure l and b?

Solution

▶ (i) If $f = l^2$, $df/dl = 2l$ so $(\Delta f)^2 = (2l\Delta l)^2$ and $\Delta(l^2) = 2l\Delta l$.
(ii) Similarly,

▶ $$\Delta(b^2) = 2b\Delta b.$$

(iii) The absolute error in $l^2 + b^2$ is obtained by adding these two terms in quadrature, i.e.

▶ $$\Delta(l^2 + b^2) = \sqrt{(4l^2[\Delta l]^2 + 4b^2[\Delta b]^2)}.$$

(iv) To find the absolute error in $(l^2 + b^2)^{1/2}$, we can proceed by noting that the *fractional* error in $(l^2 + b^2)^{1/2}$ is half the fractional error in $(l^2 + b^2)$, i.e.

$$\frac{\Delta(\sqrt{(l^2 + b^2)})}{\sqrt{(l^2 + b^2)}} = \frac{\sqrt{(4l^2(\Delta l)^2 + 4b^2(\Delta b)^2)}}{2(l^2 + b^2)}.$$

Multiplying both sides by $(l^2 + b^2)^{1/2}$ gives

▶ $$\Delta(\sqrt{(l^2 + b^2)}) = \sqrt{\frac{l^2(\Delta l)^2 + b^2(\Delta b)^2}{l^2 + b^2}}.$$

[Alternatively, we could have derived this result by writing $f = (l^2 + b^2)^{1/2}$ and using the general result $(\Delta f)^2 = (\partial f/\partial l)^2(\Delta l)^2 + (\partial f/\partial b)^2(\Delta b)^2$.]

We can see that this is reasonable by noting that if (say) $l \gg b$, it tends to Δl as expected.

▶ (v) The absolute error in $((l^2 + b^2)/12)^{1/2}$ is clearly equal to the expression derived in (iv), divided by $\sqrt{12}$.

If we require to determine k to an accuracy of 1 part in 10^5, we need to arrange that $\sqrt{(l^2 + b^2)}$ is determined to the same (fractional) accuracy. Since $l \approx 1$ m and $b \approx 0.025$ m, $\sqrt{(l^2 + b^2)}$ will be very close to 1 m, so by our result in part (iv), we require that $(l^2(\Delta l)^2 + b^2(\Delta b)^2)^{1/2}$ should be 10^{-5} m. If we assume that the errors in l and in b contribute equally to the overall error, this implies that $(l\Delta l)^2$ and $(b\Delta b)^2$ are each equal to 0.5×10^{-10} m^2, which would imply $\Delta l = 7\,\mu$m and $\Delta b = 0.3$ mm. Even if all the error is contributed by Δl, it will still be necessary to determine l to 10 μm.

▶ Thus our conclusion is that l must be determined to an accuracy of about 7 μm, and b to about 0.3 mm.

Problem 12

Calculate the moment of inertia I of a rectangular bar of density ρ, length a, width b and thickness t, about an axis through the centre of gravity, and estimate the error, given

$$I = \frac{\rho abt(a^2 + b^2)}{12} \tag{1}$$

and

$$\rho = 8.12 \pm 0.03 \text{ Mg m}^{-3},$$
$$a = 0.320 \pm 0.002 \text{ m},$$
$$b = 20 \pm 1 \text{ mm},$$
$$t = 5 \pm 0.3 \text{ mm}.$$

Solution

Straightforward substitution of the numerical values gives $I = 2.226 \times 10^{-3}$ kg m^2. The error calculation is not so simple, however, because the error in ab is not independent of the error in $(a^2 + b^2)$. It is probably easiest to use the standard result from error analysis:

$$(\Delta I)^2 = \left(\frac{\partial I}{\partial \rho}\right)^2 (\Delta \rho)^2 + \left(\frac{\partial I}{\partial a}\right)^2 (\Delta a)^2 + \left(\frac{\partial I}{\partial b}\right)^2 (\Delta b)^2$$
$$+ \left(\frac{\partial I}{\partial t}\right)^2 (\Delta t)^2,$$

where Δx is the error in x. Differentiating (1) with respect to ρ gives

$$\frac{\partial I}{\partial \rho} = \frac{abt(a^2 + b^2)}{12}$$

and substituting the values gives a numerical value of 2.74×10^{-7}. The contribution to $(\Delta I)^2$ from the error in ρ is thus $(2.74 \times 10^{-7} \times 30)^2 = 6.76 \times 10^{-11}$. Repeating the analysis for the effect of Δa, we have

$$\frac{\partial I}{\partial a} = \frac{\rho bt(a^2 + b^2)}{12} + \frac{\rho a^2 bt}{6},$$

which has a numerical value of 2.08×10^{-2}. The contribution to $(\Delta I)^2$ from the error in a is thus $(2.08 \times 10^{-2} \times 0.002)^2 = 1.73 \times 10^{-9}$. Similarly for Δb we have

$$\frac{\partial I}{\partial b} = \frac{\rho a t (a^2 + b^2)}{12} + \frac{\rho a b^2 t}{6}$$

$= 1.12 \times 10^{-1}$, giving a contribution to $(\Delta I)^2$ of $(1.12 \times 10^{-1} \times 0.001)^2 = 1.26 \times 10^{-8}$. Finally, we have

$$\frac{\partial I}{\partial t} = \frac{\rho a b (a^2 + b^2)}{12}$$

$= 4.45 \times 10^{-1}$, giving a contribution to $(\Delta I)^2$ of $(4.45 \times 10^{-1} \times 0.0003)^2 = 1.78 \times 10^{-8}$. Thus

$$(\Delta I)^2 = 6.76 \times 10^{-11} + 1.73 \times 10^{-9} + 1.26 \times 10^{-8} + 1.78 \times 10^{-8}$$
$$= 3.22 \times 10^{-8},$$

▶ so $\Delta I = 1.8 \times 10^{-4}$, and $I = (2.2 \pm 0.2) \times 10^{-3} \text{ kg m}^2$.

Problem 13

In an experiment to count decays from a radioactive sample 10 counts are registered on average in a 100 second interval. Use the Poisson distribution to estimate the probability of detecting
   (a) 8 counts in a 100 second interval, and
   (b) 2 counts in a 10 second interval.
When the experiment was run for one day, 8000 counts were detected. Explain why this is an unlikely result if the above average is correct and Poisson statistics are applicable.

Solution

The Poisson distribution can be written as

$$p(n) = \frac{\mu^n \exp(-\mu)}{n!},$$

where $p(n)$ is the probability of observing n events when the expectation value of n is μ.

(a) We know that for a 100 second interval $\mu = 10$, so

$$p(8) = \frac{10^8 \exp(-10)}{8!},$$

▶ which is evaluated as 0.113.

(b) For a 10 second interval we expect $\mu = 1$, so

$$p(2) = \frac{1^2 \exp(-1)}{2!}$$

▶ which is evaluated as 0.184.

Since one day contains 86 400 seconds, we expect $\mu = 8640$, so the observed number of 8000 events is lower than expected. In order to assess whether such a deviation is surprising, we recall that for large values of μ the Poisson distribution is very close to a Gaussian (normal) distribution with mean μ and standard deviation $\sqrt{\mu}$. The observed deviation from the expected value is thus $640/\sqrt{8640} = 6.9$ standard deviations, and we know that the probability of observing such a large deviation is extremely low.

[One possible explanation might be that the sample has decayed significantly during the course of the day. Let us estimate the decay constant which would be necessary in this case:

The count rate dN/dt will vary with time as

$$\frac{dN}{dt} = N_0' \exp(-\lambda t),$$

where N_0' is the count rate at time zero and λ is the decay constant. The total number of counts detected between time zero and time T is found by integrating dN:

$$N = N_0' \int_0^T \exp(-\lambda t) \, dt = \frac{N_0'}{\lambda}(1 - \exp[-\lambda T]).$$

If we put $x = \lambda T$, where $T = 1$ day $= 86\,400$ s, and $N_0' = 0.1 \,\mathrm{s}^{-1}$, we obtain the expression

$$1 - \exp(-x) = 0.925\,93x.$$

To obtain an approximate solution of this equation, we can expand $\exp(-x)$ as a power series in x to give

$$x - x^2/2 + x^3/6 - \ldots = 0.925\,93x,$$

hence

$$0.074\,07x - x^2/2 + x^3/6 \ldots = 0.$$

Eliminating the possibility that $x = 0$ leaves us with a quadratic equation in x:

$$x^2/6 - x/2 + 0.074\,07 = 0,$$

which can be solved to give $x \approx 0.1563$. (The other solution, $x \approx 2.84$, can be rejected by substitution into the equation $1 - \exp(-x) = 0.925\,93x$.) Thus we estimate the decay constant λ as $0.1563/(86\,400\text{ s}) = 1.81 \times 10^{-6}\,\text{s}^{-1}$, corresponding to a half life of $\ln 2/\lambda \approx 4.4$ days.]

Classical mechanics and dynamics

Problem 14

Show that the maximum range of a projectile of fixed initial speed is obtained when it is launched at an angle of 45° to the horizontal. Ignore the effects of air resistance.

The initial speed of a bullet fired from a rifle is 630 m s^{-1}. The rifle is fired at the centre of a target 700 m away at the same level as the target. In order to hit the target the rifle must be aimed at a point above the target. How far above the centre of the target must the rifle be aimed? What will be the maximum height reached by the bullet along its trajectory?

Solution

Figure 5 shows the trajectory of the projectile described in a Cartesian coordinate system. If the projectile is launched at time $t = 0$ with speed v and at an angle θ to the horizontal, its x-coordinate (in the horizontal direction) at time t is

$$x = vt \cos \theta$$

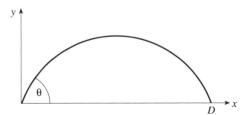

Figure 5

and its y-coordinate (in the vertical direction) is

$$y = vt \sin \theta - \frac{1}{2}gt^2.$$

When the projectile strikes the ground, $y = 0$ so $t = (2v \sin \theta)/g$. The range D is thus given by

$$D = v \cos \theta \frac{2v \sin \theta}{g} = \frac{2v^2 \sin \theta \cos \theta}{g} = \frac{v^2 \sin (2\theta)}{g}.$$

Since $\sin x$ takes its maximum value when $x - 90°$, the maximum range is achieved when $2\theta = 90°$ so $\theta = 45°$ as required.

Figure 6 shows the trajectory of the rifle bullet.

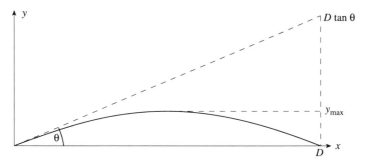

Figure 6

For the bullet, we require that $D = 700$ m when $v = 630 \text{ m s}^{-1}$. Taking $g = 9.81 \text{ m s}^{-2}$ gives $\theta = 0.496°$. [There is another solution in the range $0° < \theta < 90°$. It is $89.504°$, and although it is valid physically, common sense suggests that one would not aim a rifle almost vertically in order to hit a target 700 m away in a horizontal direction!] The rifle must therefore be aimed at a point $D \tan \theta = 6.1$ m above the centre of the target. The bullet attains its maximum height when $dy/dt = 0$, and since

$$dy/dt = v \sin \theta - gt,$$

this happens at time $t = (v \sin \theta)/g$. Substituting this value of t into the expression for y gives

$$y_{\max} = \frac{v^2 \sin^2 \theta}{2g},$$

and substitution of our values for v and θ gives $y_{\max} = 1.52$ m.

Problem 15

A projectile is fired uphill over ground which slopes at an angle α to the horizontal. Find the direction in which it should be aimed to achieve the maximum range.

Solution

As in problem 14, we will set up a coordinate system in which the x-axis is horizontal, the y-axis is vertical, and the origin is located at the point from which the projectile is launched (see figure 7).

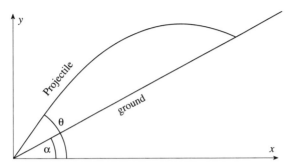

Figure 7

The coordinates of the projectile are again given by

$$x = vt \cos \theta,$$
$$y = vt \sin \theta - gt^2/2,$$

where θ is the angle the projectile makes with the horizontal at the instant of launch. However, when the projectile strikes the ground its coordinates must satisfy the equation

$$y = x \tan \alpha,$$

rather than $y = 0$ as in problem 14. Making the substitution, we obtain

$$vt \sin \theta - \frac{1}{2}gt^2 = vt \cos \theta \tan \alpha,$$

which can be rearranged to give

$$t = \frac{2v}{g}(\sin \theta - \cos \theta \tan \alpha).$$

Substituting this into our expression for x gives

$$x = \frac{2v^2}{g}(\cos \theta \sin \theta - \cos^2 \theta \tan \alpha).$$

Now the range will be a maximum when x is maximum, so we can differentiate this expression with respect to θ and set $dx/d\theta = 0$ to find the condition for the maximum range:

$$\frac{dx}{d\theta} = \frac{2v^2}{g}(\cos^2 \theta - \sin^2 \theta + 2 \cos \theta \sin \theta \tan \alpha).$$

Setting this equal to zero and recalling that $\cos^2 \theta - \sin^2 \theta = \cos(2\theta)$ and that $2\cos \theta \sin \theta = \sin(2\theta)$ gives

$$\tan(2\theta) = \frac{-1}{\tan \alpha}.$$

We have done what we were asked, since we have produced an expression relating θ to α such that the range will be maximum, but the expression can be simplified further. Since

$$-1/\tan \alpha = \tan(\alpha - \pi/2 + n\pi),$$

where n is an integer, we obtain the expression

$$\theta = \alpha/2 - \pi/4 + n\pi/2.$$

n must be 1 otherwise θ would not lie within the range 0 to $\pi/2$, so we can finally write our solution as $\theta = \alpha/2 + \pi/4$. [This is in fact the direction that bisects the angle between the slope and the vertical.]

Problem 16

Rockets are propelled by the ejection of the products of the combustion of fuel. Consider a rocket of total mass m_1 to be travelling at speed v_1 in a region of space where gravitational forces are negligible. Suppose that the combustion products are ejected at a constant speed v_r relative to the rocket. Show that a fuel 'burn' which reduces the total mass of the rocket to m_2 results in an increase in the speed of the rocket to v_2, such that

$$v_2 - v_1 = v_r \ln \frac{m_1}{m_2}.$$

Suppose that 2.1×10^6 kg of fuel are consumed during a 'burn' lasting 1.5×10^2 s. Given that there is a constant force on the rocket of 3.4×10^7 N during this 'burn', calculate v_r. Hence calculate the increase in speed resulting from the 'burn' if m_1 is 2.8×10^6 kg.

What is the initial vertical acceleration that can be imparted to this rocket when it is launched from Earth, if the initial mass is 2.8×10^6 kg?

Solution

It is probably easiest to consider the acceleration of the rocket in its *instantaneous rest frame*, in which the rocket begins from rest at time t. At time $t + dt$ the rocket has acquired a velocity dv in the forward direction, and its mass has changed from m to $m + dm$. Since the mass has in fact decreased, dm is negative. The mass of combustion products emitted in this time is thus $-dm$ (a positive quantity), and this mass is travelling backwards at velocity v_r, as shown in figure 8.

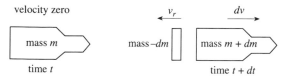

velocity zero

mass m

time t

v_r

dv

mass $-dm$

mass $m + dm$

time $t + dt$

Figure 8

In this frame, the total momentum at time t is zero, so the momentum at time $t + dt$ must also be zero since no external forces act upon the system. Thus

$$-v_r dm = (m + dm)dv.$$

Ignoring the second-order term $dm\,dv$, we obtain the expression

$$dv = -v_r \frac{dm}{m}$$

for the increase in the rocket's velocity when its mass changes by dm. We can now integrate this expression to obtain

$$v_2 - v_1 = -v_r \int_{m_1}^{m_2} \frac{dm}{m} = v_r \ln \frac{m_1}{m_2}$$

as required.

The acceleration of the rocket is dv/dt, and our expression for dv shows that this can be written as $-v_r\,(dm/dt)/m$. Thus the force F acting on the rocket is given by

$$F = -v_r \frac{dm}{dt}.$$

If we take $dm/dt = -(2.1 \times 10^6 \text{ kg})/(1.5 \times 10^2 \text{ s}) = -1.4 \times 10^4 \text{ kg s}^{-1}$ and $F = 3.4 \times 10^7$ N, we find that $v_r = 2.43 \times 10^3 \text{ m s}^{-1}$. The increase in speed resulting from the 'burn' is

$$\begin{aligned} \Delta v &= v_r \ln(m_1/m_2) \\ &= 2.4 \times 10^3 \ln(2.8/0.7) \text{ m s}^{-1} \\ &= 3.4 \times 10^3 \text{ m s}^{-1}. \end{aligned}$$

Assuming that the thrust on the rocket is 3.4×10^7 N during launch, and that the initial mass is 2.8×10^6 kg, the accelerating force is equal to the thrust minus the weight, or $(3.4 \times 10^7 - 9.8 \times 2.8 \times 10^6) \text{ N} = 6.6 \times 10^6$ N. The vertical acceleration is thus $(6.6 \times 10^6)/(2.8 \times 10^6) \text{ m s}^{-2} = 2.3 \text{ m s}^{-2}$.

Problem 17

A chlorine molecule with an initial velocity of $600\,\mathrm{m\,s^{-1}}$ absorbs a photon of wavelength 350 nm and is then dissociated into two chlorine atoms. One of the atoms is detected moving perpendicular to the initial direction of the molecule and having a velocity of $1600\,\mathrm{m\,s^{-1}}$. Calculate the binding energy of the molecule. [Neglect the momentum of the absorbed photon. The relative atomic mass of chlorine is 35.]

Solution

Let us put v for the initial speed and u for the speed of the atom which moves perpendicularly (as shown in figure 9), and resolve the motion of the non-perpendicularly moving atom into components parallel and perpendicular to the initial direction. Since the molecule initially has no momentum in the perpendicular direction, the perpendicular components of the velocities of the two atoms must be equal and opposite, with magnitude u. Conservation of the parallel component of momentum shows that the non-perpendicularly moving atom must have a component of velocity $2v$ in the parallel direction (see figure 10).

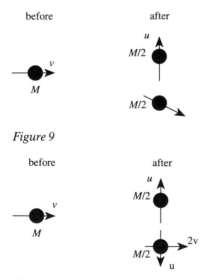

Figure 9

Figure 10

The final kinetic energy is

$$\frac{1}{2}\frac{M}{2}u^2 + \frac{1}{2}\frac{M}{2}(u^2 + [2v]^2) = \frac{1}{2}Mu^2 + Mv^2.$$

The initial kinetic energy was $Mv^2/2$, to which was added the photon energy hc/λ (where h is Planck's constant, c is the speed of light and λ is the wavelength of the photon). The binding energy E_b was required to break the Cl–Cl bond, so conservation of energy gives

$$\frac{1}{2}Mv^2 + \frac{hc}{\lambda} - E_b = \frac{1}{2}Mu^2 + Mv^2.$$

Rearranging:

$$E_b = \frac{hc}{\lambda} - \frac{1}{2}Mv^2 - \frac{1}{2}Mu^2.$$

Substituting $\lambda = 350$ nm, $M = 70 \times 1.66 \times 10^{-27}$ kg, $v = 600\,\mathrm{m\,s^{-1}}$ and
$u = 1600\,\mathrm{m\,s^{-1}}$ gives $E_b = 4.0 \times 10^{-19}$ J. [A physicist would probably check that this is reasonable by converting it to 2.5 electron volts, and a chemist would probably prefer to express it as 240 kJ mol^{-1}.]

[We ought to check that it is reasonable to ignore the momentum of the photon. Clearly for this to be so, the ratio of the photon's momentum to the momentum of the chlorine molecule should be much less than unity. The ratio is $h/\lambda Mv$, which has a value of 3×10^{-5}, so the assumption is a safe one.]

Problem 18

A body of mass m_1 collides elastically with a stationary mass m_2 and after collision the bodies move making angles θ_1 and θ_2 with the original direction of m_1. By considering events in the centre of mass frame, or otherwise, show that
  (a) if $m_1 = m_2$, $\theta_1 = \pi/2 - \theta_2$,
  (b) if $m_1 > m_2$, the maximum value of θ_1 is given by
    $\sin \theta_{\max} = m_2/m_1$,
  (c) if $m_1 \ll m_2$, $\theta_1 \approx \pi - 2\theta_2$.

Solution

Before the collision, the appearance of the system in the original (laboratory) frame is as shown in figure 11.

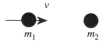

Figure 11

The total momentum of the system is m_1v and the total mass is $m_1 + m_2$, so the velocity of the centre of mass frame (the zero momentum frame) is

$$\frac{m_1v}{m_1 + m_2}$$

to the right. The velocities of the two masses in the centre of mass frame are found by subtracting this velocity from their velocities in the laboratory frame. Thus the initial velocity of mass m_1 in the centre of mass frame is $m_2v/(m_1 + m_2)$, and the initial velocity of mass m_2 is $-m_1v/(m_1 + m_2)$. We can draw a diagram in the centre of mass frame, as shown in figure 12.

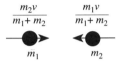

Figure 12

Since the collision is elastic, kinetic energy is conserved, and since no external forces act on the system, momentum is conserved. In the centre of mass frame the total momentum is, by definition, zero, so the only way in which these two conditions can be satisfied is for the two bodies to retain their speeds in the centre of mass frame. Their directions may change, but the velocities in the centre of mass frame must be in opposite directions to conserve momentum. Thus the situation after the collision, as seen in the centre of mass frame, is as shown in figure 13.

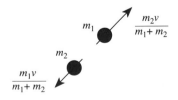

Figure 13

We can now convert back to the laboratory frame by adding a velocity $m_1v/(m_1 + m_2)$ to the right, as shown in figure 14.

(a) If $m_1 = m_2$, the lines OD, DA, AP, PB, BC, CO and OP are all of equal length. Triangles ODA and OPA are congruent so that angle $DOA = \theta_1$, and triangles OCB and OPB are congruent so that angle

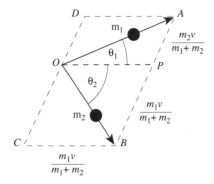

Figure 14

▶ $COB = \theta_2$. Thus $2\theta_1 + 2\theta_2 = \pi$, and $\theta_1 = \pi/2 - \theta_2$. (I.e. the two masses move at right angles to each other after the collision.)

(b) To find the maximum value of θ_1, it is helpful to extract part of figure 14, shown in figure 15.

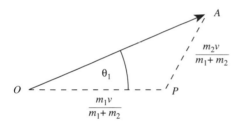

Figure 15

The maximum value of θ_1 occurs when OA is perpendicular to AP, so

▶ $$\sin \theta_{\max} = \frac{m_2}{m_1}.$$

(c) Triangles OCB and OPB are congruent so angle $COB = \theta_2$. If $m_1 \ll m_2$, angle DOA is very small (because $DA \ll DO$). Thus
▶ $\pi \approx \theta_1 + 2\theta_2$, and hence $\theta_1 \approx \pi - 2\theta_2$.

Problem 19

A body of mass m is moving in one dimension under the influence of a conservative force with a potential energy given by $U(x)$. Show that the body, when displaced slightly from a point of stable equilibrium at

$x = x_0$, will experience a restoring force proportional to its displacement, the force constant being

$$\left[\frac{d^2 U}{dx^2}\right]_{x=x_0}.$$

Suppose the potential energy has the form

$$U(x) = \frac{-cx}{x^2 + a^2},$$

where c and a are positive constants. Sketch this potential and the force resulting from it. Find the position of stable equilibrium and calculate the angular frequency of small oscillations about this position.

Solution

The force F is given by $-dU/dx$, and if x_0 is a position of equilibrium the force must be zero at this position. To find the force at a nearby position x we can use a Taylor expansion:

$$F(x) = F(x_0) + \frac{dF}{dx}\Delta x + \text{terms in higher powers of } \Delta x,$$

where $\Delta x = x - x_0$ and the differential is evaluated at $x = x_0$. Since $F(x_0)$ is zero, the force in the direction of increasing x is proportional to the displacement Δx. The restoring force (in the direction of decreasing

▶ x) is $-F$, so the force constant is $-dF/dx = +d^2 U/dx^2$ as required.

We can differentiate the potential

$$U(x) = \frac{-cx}{x^2 + a^2}$$

to obtain

$$\frac{dU}{dx} = -\frac{c}{x^2 + a^2} + \frac{2cx^2}{(x^2 + a^2)^2},$$

so the force is given by the negative of this:

$$F(x) = \frac{c}{x^2 + a^2} - \frac{2cx^2}{(x^2 + a^2)^2}.$$

Sketches of these two functions are shown in figure 16.

Although the graph of $F(x)$ is symmetric, the position $x = -x_0$ is not a stable equilibrium because if the body is moved towards $x = 0$ it experiences a positive force (i.e. in the direction of motion). To find the

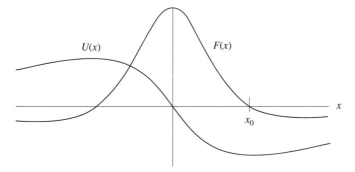

Figure 16

position x_0 of stable equilibrium we set $F(x_0) = 0$, which gives

$$\frac{c}{x_0^2 + a^2} = \frac{2cx_0^2}{(x_0^2 + a^2)^2}.$$

Hence

$$2x_0^2 = x_0^2 + a^2,$$

so that $x_0 = \pm a$. By inspecting our sketch, we see that the equilibrium
▶ position must be given by $x_0 = a$.

To find the angular frequency of small oscillations we need to know the
restoring force constant, which is given by d^2U/dx^2 at $x = x_0$.
Differentiating dU/dx gives

$$\frac{d^2U}{dx^2} = \frac{6cx}{(x^2 + a^2)^2} - \frac{8cx^3}{(x^2 + a^2)^3},$$

and putting $x = x_0 = a$ gives $d^2U/dx^2 = c/2a^3$. We know that a body of
mass m subject to a restoring force $k\Delta x$ will oscillate with an angular
frequency ω given by $\omega^2 = k/m$, so the angular frequency of small
oscillations is

▶ $$\omega = \sqrt{\frac{c}{2a^3 m}}.$$

Problem 20

A ball of mass 0.5 kg attached to a light inextensible string rotates in a
vertical circle of radius 0.75 m, such that it has a speed of $5\,\mathrm{m\,s^{-1}}$ when
the string is horizontal. Calculate the speed of the ball and the tension in

the string at the highest and lowest points on its circular path. Evaluate the work done by the Earth's gravitational force and by the tension in the string as the ball moves from its highest to its lowest point.

Solution

Let us put v for the speed of the ball when the string is horizontal, w for its speed at the top of the circle, and u for its speed at the bottom of the circle, as shown in figure 17.

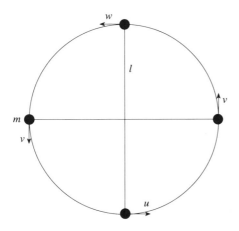

Figure 17

The problem reminds us that the two possible mechanisms for doing work on the ball are the gravitational force and the tension in the string. However, the tension can in fact do *no* work on the ball since the force is always perpendicular to the velocity. Thus we need consider only the interchange of kinetic and potential energy:

$$\frac{1}{2}mu^2 - \frac{1}{2}mv^2 = \frac{1}{2}mv^2 - \frac{1}{2}mw^2 = mgl,$$

so

$$u^2 - v^2 = v^2 - w^2 = 2gl.$$

▶   Taking $l = 0.75$ m and $v = 5$ m s^{-1} gives $w = 3.2$ m s^{-1} and $u = 6.3$ m s^{-1}.
At the highest point, the centripetal force is $mw^2/l = 6.86$ N downwards. mg of this is supplied by the weight of the ball, so the tension
▶   in the string is $6.86 - 0.5 \times 9.81 = 2.0$ N.
At the lowest point, the centripetal force is $mu^2/l = 26.48$ N upwards.

However, the tension in the string must also support the ball's weight, so
▶ the tension must be $26.48 + 0.5 \times 9.81 = 31.4$ N.

The work done by gravity as the ball moves from the lowest point to
▶ the highest is $2\,mgl = 7.4$ J.

Problem 21

A car of mass m travelling at speed v moves on a horizontal track so that
the centre of mass describes a circle of radius r. Show that the limiting
speed beyond which the car will overturn is given by

$$v^2 = \frac{gra}{h},$$

where $2a$ is the separation of the inner and outer wheels and h is the
height of the centre of mass above the ground.

Solution

Let us assume that the car is travelling into the page and is turning
towards the left. It must be subject to a centripctal force mv^2/r to the
left, provided by frictional forces acting on the tyres. We can draw a rear
view of the system, as in figure 18.

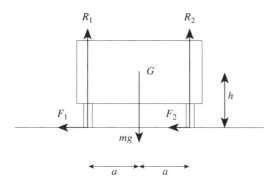

Figure 18

Resolving vertically,
$$R_1 + R_2 = mg. \tag{1}$$

Resolving horizontally,
$$F_1 + F_2 = \frac{mv^2}{r}. \tag{2}$$

Its angular velocity ω is v/r and its moment of inertia I is $mr^2/2$ (see problem 23). Thus its total kinetic energy is

$$\frac{1}{2}mv^2 + \frac{1}{2}I\omega^2 = \frac{1}{2}mv^2 + \frac{1}{2}\frac{1}{2}mr^2\frac{v^2}{r^2} = \frac{3}{4}mv^2.$$

It is independent of the coin's radius (which was not given in the problem). Substituting $m = 0.010$ kg and $v = 0.06$ m s^{-1} gives a kinetic energy of 27 μJ.

▶

Problem 23

Show that the moment of inertia of a uniform disc of mass M and radius R about an axis through its centre and perpendicular to its surface is equal to $MR^2/2$.

A record turntable is accelerated at a constant rate from 0 to $33\frac{1}{3}$ revolutions per minute in two seconds. It is a uniform disc of mass 1.5 kg and radius 13 cm. What torque is required to provide this acceleration and what is the angular momentum of the turntable at its final speed?

A mass of 0.2 kg is dropped vertically and sticks to the freely rotating turntable at a distance of 10 cm from its centre. What is the angular velocity of the turntable after the mass has been added?

Solution

The moment of inertia of a body is given by

$$I = \sum_i m_i r_i^2,$$

where the body is considered to be made up of point masses such that the ith mass is m_i and it is at a perpendicular distance r_i from the axis of rotation. For the disc, we can calculate its moment of inertia by adding the contributions from annuli of radius r and width dr, since all the mass within such an annulus is at the same distance from the axis of rotation. Let us put σ for the mass per unit area of the disc. Since the area of the annulus is $2\pi r\,dr$, its mass is $2\pi r\sigma\,dr$ and its moment of inertia is $2\pi r^3\sigma\,dr$. We can integrate this expression from $r = 0$ to $r = R$ to find the total moment of inertia:

$$I = 2\pi\sigma\int_0^R r^3\,dr = \frac{\pi\sigma R^4}{2}.$$

The total mass of the disc is $M = \sigma \pi R^2$ so we can use this expression to eliminate σ, giving

▶ $$I = \frac{MR^2}{2}$$

as required.

The torque G required to give a body of moment of inertia I an angular acceleration $d\omega/dt$ is $I d\omega/dt$. Since the angular acceleration of the turntable is uniform, we can write

$$G = \frac{I\omega_{\text{final}}}{T},$$

where ω_{final} is the final angular velocity and T is the time taken to reach this value. From the data,

$$I = \frac{1}{2} 1.5 \, (0.13)^2 = 0.0127 \text{ kg m}^2$$

and

$$\omega_{\text{final}} = \frac{33\frac{1}{3} 2\pi}{60} = 3.491 \text{ s}^{-1},$$

▶
▶ so taking $T = 2\,\text{s}$ gives $G = 0.0221\,\text{N m}$. The angular momentum $J = I\omega_{\text{final}} = 0.0442\text{ kg m s}^{-1}$.

When the mass is dropped onto the turntable and sticks to it, it increases the moment of inertia by an amount mr^2, where m is the mass and r is the distance from the axis of rotation (see figure 21). Taking $m = 0.2$ kg and $r = 0.1$ m gives a value of 0.002 kg m^2 for the increase in I, so its new value is 0.0147 kg m^2. We are told that the turntable is freely

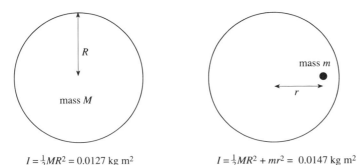

$I = \frac{1}{2}MR^2 = 0.0127$ kg m^2 $I = \frac{1}{2}MR^2 + mr^2 = 0.0147$ kg m^2

Figure 21

and l for the length of the equivalent rod. Since the moment of inertia of a rod about its centre is

$$\frac{1}{12}ml^2,$$

the skater's total moment of inertia when her arms are outstretched is

$$I_1 = Mk^2 + \frac{1}{12}ml^2.$$

When she pulls her arms in, we assume that they lie along her rotation axis so they will contribute nothing to her moment of inertia which thus becomes

$$I_2 = Mk^2$$

(see figure 23). Since no external torque acts on the skater, her angular momentum must be conserved, so if we put ω_1 and ω_2 respectively for her initial and final angular velocities, we must have

$$I_1\omega_1 = I_2\omega_2$$

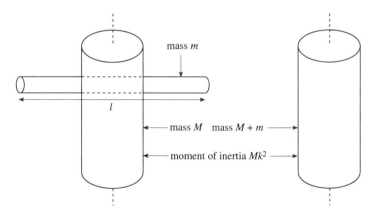

Figure 23

and hence

$$\frac{I_1}{I_2} = \frac{\omega_2}{\omega_1} = 3.$$

Thus, using our expressions for I_1 and I_2,

$$\frac{1}{12}ml^2 = 2Mk^2,$$

so

$$k^2 = \frac{ml^2}{24M}.$$

▶ Putting $m = 5$ kg, $M = 70$ kg and $l = 1.4$ m gives $k = 0.076$ m. [We can check whether this is a reasonable answer by approximating the skater's body to a uniform vertical cylinder of radius r. Such a cylinder has a radius of gyration of $r/\sqrt{2}$, hence the radius must be about 11 cm and the circumference about 70 cm. This is clearly a reasonable figure for the 'average' circumference of an ice skater.]

We can see that the skater's kinetic energy must have increased as follows. Her kinetic energy E is given by $I\omega^2/2$, where I is her moment of inertia and ω is her angular velocity. Her angular momentum J is given by $I\omega$, and this is constant since no external torque acts. Thus we can rewrite the kinetic energy as $E = J^2/2I$, which shows that if I decreases, E must increase.

Problem 26

Find the ratio of the height h of a cushion on a snooker table to the radius r of a ball as shown in figure 24, such that when the ball hits the cushion with a pure rolling motion it rebounds with a pure rolling motion. (Assume that the force exerted on the ball by the cushion is horizontal during the impact and that the ball hits the cushion normally.)

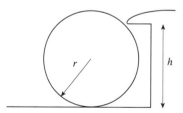

Figure 24

Solution

Lét the ball's mass be m and its moment of inertia I, and suppose that it is initially rolling to the right, as shown in figure 25, with speed u. We will also suppose that it rebounds to the left at speed v, so that the change in momentum is $m(v + u)$. The impulse P exerted on the ball by the cushion is thus $m(v + u)$.

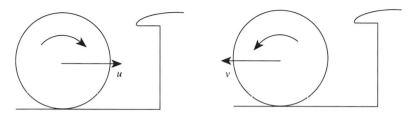

Figure 25

   The line of action of this impulse passes at a distance $h - r$ from the centre of the ball, so that ball's angular momentum is changed by an amount $P(h - r)$ in an anticlockwise sense. The ball initially has a clockwise angular velocity u/r (since it is rolling without slipping), so its initial angular momentum is Iu/r. After the collision, if the ball is still rolling without slipping its angular velocity must be v/r anticlockwise, so the change of angular momentum is

$$\frac{I(u + v)}{r}.$$

Thus we have

$$\frac{I(u + v)}{r} = P(h - r) = m(u + v)(h - r),$$

hence

$$r(h - r) = \frac{I}{m}.$$

Now for a solid, uniform sphere of radius r,

$$\frac{I}{m} = \frac{2r^2}{5},$$

so

$$hr - r^2 = 2r^2/5,$$

▶ hence $h = 7r/5$.

Problem 27

A thin uniform rectangular plate of length a, width b and mass m has a moment of inertia

$$\frac{1}{12}m(a^2 + b^2)$$

about an axis through its centre and perpendicular to its plane. Identify
the principal axes of inertia and give the values of the principal moments
of inertia.

What is the moment of inertia of the plate about an axis in the plane of
the plate and forming a diagonal of the rectangle?

If the plate is made to rotate about the diagonal, show that the angle
between the angular velocity and the angular momentum is

$$\arctan \frac{a}{b} - \arctan \frac{b}{a}.$$

Solution

The three principal axes of a cuboid are the three axes of symmetry,
normal to the faces of the cuboid. Since a thin rectangular plate is a
special case of a cuboid, one of the principal axes is normal to the plate,
and the other two are in the plane of the plate and parallel to the edges.
All three principal axes pass through the centre of the plate, as shown in
figure 26.

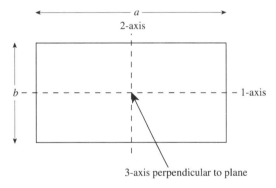

Figure 26

The perpendicular axes theorem for a lamina (thin plate) states that

$$I_1 + I_2 = I_3$$

and since we expect I_1 to be proportional to mb^2 and I_2 to be
proportional to ma^2, it clearly follows that

$$I_1 = \frac{1}{12}mb^2 \text{ and } I_2 = \frac{1}{12}ma^2.$$

[These are identical to the results for rods of length b and a about the
centres, and we could of course have derived them from first principles.]

To find the moment of inertia about a diagonal, it is easiest to consider two sets of $x-y$ coordinates, one aligned with the rectangle and one aligned with its diagonal, as shown in figure 27.

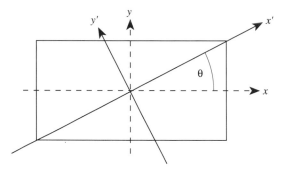

Figure 27

The moment of inertia about the x'-axis is given by

$$I_{x'} = \sum m_i y'^2_i,$$

where m_i is the mass of an element of the plate, and the sum is carried out over the whole plate. Now

$$y' = y \cos \theta - x \sin \theta,$$

so

$$I_{x'} = \sum m_i y^2_i \cos^2 \theta + \sum m_i x^2_i \sin^2 \theta - 2\sum m_i x_i y_i \cos \theta \sin \theta.$$

But we also know that

$$\sum m_i x^2_i = I_2 = \frac{1}{12} m a^2,$$

$$\sum m_i y^2_i = I_1 = \frac{1}{12} m b^2,$$

$$\sum m_i x_i y_i = 0 \quad \text{(by symmetry)},$$

so that

$$I_{x'} = \frac{m}{12} (a^2 \sin^2 \theta + b^2 \cos^2 \theta).$$

We also know that

$$\cos^2 \theta = \frac{a^2}{a^2 + b^2} \quad \text{and} \quad \sin^2 \theta = \frac{b^2}{a^2 + b^2},$$

so we finally obtain

▶ $$I_{x'} = \frac{1}{6} m \frac{a^2 b^2}{a^2 + b^2}.$$

If the plate is rotated at angular velocity ω about its diagonal, the angular velocity has a component $\omega \cos \theta$ in the x-direction and a component $\omega \sin \theta$ in the y-direction. The angular momentum thus has a component $I_1 \omega \cos \theta$ in the x-direction and a component $I_2 \omega \sin \theta$ in the y-direction, so the angle between the direction of the angular momentum axis and the x-axis is

$$\arctan \frac{I_2 \omega \sin \theta}{I_1 \omega \cos \theta} = \arctan \frac{a^2 \sin \theta}{b^2 \cos \theta} = \arctan \frac{a}{b}.$$

The angle between the angular velocity vector and the x-axis is $\theta = \arctan(b/a)$, so the angle between the two vectors is

▶ $$\arctan \frac{a}{b} - \arctan \frac{b}{a}$$

as required.

Problem 28

A cotton reel is made up of a hub of radius a and two end caps of radius b. The mass of the complete reel is m and its moment of inertia about its longitudinal axis is I. The reel rests on a perfectly rough table (so that only rolling motion is possible) and a tension T is applied to the free end of the cotton wrapped around the hub as shown in figure 28. In what direction does the reel begin to move? Find the frictional force exerted by the table and the direction in which it acts.

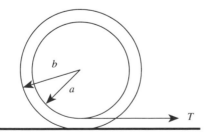

Figure 28

Solution

Let us assume that the reel moves to the right, and therefore rolls clockwise. Since the tension in the thread exerts an anticlockwise moment on the reel, the frictional force must exert a larger clockwise moment. The forces acting on the reel are thus (ignoring vertical forces acting through the centre) as shown in figure 29.

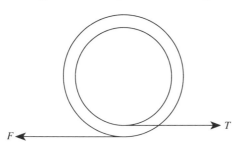

Figure 29

The net force to the right is $T - F$ so the linear acceleration dv/dt is

$$\frac{T - F}{m}.$$

The net clockwise moment is $bF - aT$ so the angular acceleration $d\omega/dt$ is

$$\frac{bF - aT}{I}.$$

Now for a pure rolling motion the velocity v and the angular velocity ω are related by $v = b\omega$, so

$$\frac{dv}{dt} = b\frac{d\omega}{dt}.$$

Thus

$$\frac{T - F}{m} = \frac{b(bF - aT)}{I}.$$

Rearranging this to obtain an expression for F in terms of T gives

$$F(I + mb^2) = T(I + mab),$$

so

$$F = T\frac{I + mab}{I + mb^2}.$$

To check that our initial assumption that the reel moves to the right was correct, we can use this result to calculate the net force to the right,

$$T - F = \frac{Tmb(b - a)}{I + mb^2},$$

and the net clockwise moment,

$$bF - aT = \frac{IT(b - a)}{I + mb^2},$$

which are both positive as we assumed.

Problem 29

An elastic spherical ball of mass M and radius a moving with velocity v strikes a rigid surface at an angle θ to the normal. Assuming it skids while in contact with the surface, the tangential frictional force being a constant fraction μ of the normal reaction force, show that
  (a) the ball is reflected at an angle ϕ to the normal where
  $|\tan \theta - \tan \phi| = 2\mu,$
  (b) the angular velocity of the rebounding ball changes by an amount

$$\frac{5\mu v}{a} \cos \theta.$$

You may assume that the component of velocity perpendicular to the surface is reversed in direction without change of magnitude.

Solution

The problem refers to a *change* in the angular velocity of the ball, so it is presumably not safe to assume that the ball is not initially rotating (and in fact if the ball were not rotating, it could not skid against the surface so the angles θ and ϕ would be identical). Let us therefore assume that it has an initial angular velocity ω_1 and a final angular velocity ω_2, as shown in figure 30.

(a) We are told that the perpendicular component of the velocity is reversed by the impact, so (with the notation of figure 30)

$$v \cos \theta = u \cos \phi. \tag{1}$$

The change in the perpendicular component of the momentum is $2Mv \cos \theta$, so this must be the perpendicular component of the impulse

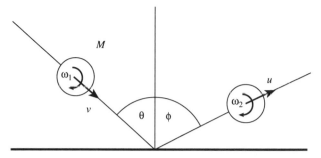

Figure 30

delivered to the ball by the surface. Since the frictional force is μ times the perpendicular force, the component of the impulse parallel to the surface must be

$$2Mv\mu\cos\theta.$$

If the ball is rotating clockwise as shown in the diagram, this impulse will be to the right, but it would be to the left if the ball were rotating anticlockwise.

The parallel component of the ball's momentum must thus change by $2Mv\mu\cos\theta$, so the parallel component of the velocity changes by $2v\mu\cos\theta$. Thus

$$u\sin\phi - v\sin\theta = 2\mu v\cos\theta. \tag{2}$$

Equations (1) and (2) give

$$\tan\phi - \tan\theta = 2\mu,$$

but if the original angular velocity were anticlockwise instead of clockwise we could obtain

$$\tan\phi - \tan\theta = -2\mu,$$

so we may put

▶ $$|\tan\theta - \tan\phi| = 2\mu.$$

(b) We have shown that the horizontal component of the impulse delivered to the ball is $2Mv\mu\cos\theta$, and since the ball's radius is a the impulsive moment (angular impulse) is $2Mv\mu a\cos\theta$. Since the impulsive moment is equal to the change in angular momentum, and the angular momentum is given by $I\omega$ where I is the moment of inertia, we must have

$$\Delta\omega = \frac{2\mu M v a \cos\theta}{I} = \frac{2\mu M v a \cos\theta}{\frac{2}{5}Ma^2} = \frac{5\mu v \cos\theta}{a}.$$

Problem 30

A pendulum is constructed from two identical uniform thin rods a and b each of length L and mass m, connected at right angles to form a 'T' by joining the centre of rod a to one end of rod b. The 'T' is then suspended from the free end of rod b and the pendulum swings in the plane of the 'T'.

(a) Calculate the moment of inertia I of the 'T' about the axis of rotation.

(b) Give expressions for the kinetic and potential energies in terms of the angle θ of inclination to the vertical of the pendulum.

(c) Derive the equation of motion of the pendulum.

(d) Show that the period of small oscillations is

$$2\pi\sqrt{\frac{17L}{18g}}.$$

Solution

(a) Figure 31 shows the 'T' suspended vertically.

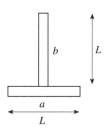

Figure 31

The moment of inertia of a thin rod of length L and mass m about its centre is $mL^2/12$, so application of the parallel axes theorem gives the moment of inertia of rod b as

$$I_b = \frac{1}{12}mL^2 + m\left(\frac{L}{2}\right)^2 = \frac{1}{3}mL^2.$$

Application of the parallel axes theorem to rod a gives its moment of inertia about the suspension point as

$$I_a = \frac{1}{12}mL^2 + mL^2 = \frac{13}{12}mL^2.$$

▶ Thus the total moment of inertia of the system is $I - (1/3 + 13/12)mL^2 = 17mL^2/12$.

(b) Figure 32 shows the 'T' at an angle θ to the vertical.

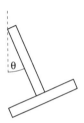

Figure 32

The kinetic energy of the system is given by

▶ $$\frac{1}{2}I\left(\frac{d\theta}{dt}\right)^2.$$

At an angle θ the centre of mass of rod b has been raised through a distance $L(1 - \cos \theta)/2$ and the centre of mass of rod a has been raised by $L(1 - \cos \theta)$, so the potential energy is

▶ $$\frac{3mgL}{2}(1 - \cos \theta)$$

with respect to the value at $\theta = 0$.

(c) The equation of motion of the pendulum can be derived using the fact that the total energy is constant:

$$\frac{17}{24}mL^2\left(\frac{d\theta}{dt}\right)^2 + \frac{3mgL}{2}(1 - \cos \theta) = \text{constant}.$$

Differentiating this expression with respect to time gives

$$\frac{17}{12}mL^2\left(\frac{d\theta}{dt}\right)\left(\frac{d^2\theta}{dt^2}\right) + \frac{3mgL}{2}\sin \theta \left(\frac{d\theta}{dt}\right) = 0,$$

which can be rearranged to give

$$\frac{d^2\theta}{dt^2} = -\frac{18}{17}\frac{g\sin\theta}{L}.$$

[Note that we could also have derived this result by using Lagrange's equations of motion.]

(d) When the oscillations are small, we may put $\sin\theta \approx \theta$ so that the equation of motion becomes

$$\frac{d^2\theta}{dt^2} = -\frac{18}{17}\frac{g}{L}\theta.$$

We recognise this as the equation of simple harmonic motion with angular frequency $(18g/17L)^{1/2}$, so the period of small oscillations is

$$2\pi\sqrt{\frac{17L}{18g}}$$

as required.

Problem 31

A uniform rod of length a is freely pivoted at one end. It is initially held horizontally and then released from rest. What is the angular velocity at the instant when the rod is vertical? When the rod is vertical it breaks at its midpoint. What is the largest angle from the vertical reached by the upper part of the rod in its subsequent motion? Describe the motion of the lower part of the rod. (Assume that no impulsive forces are generated when the rod breaks.)

Solution

Suppose the rod has mass m. In rotating from a horizontal to a vertical position, the centre of mass falls through a distance $a/2$ so the loss of potential energy is $mga/2$. Since the rod is initially at rest, its kinetic energy as it passes through the vertical position must therefore also be given by $mga/2$.

Now the moment of inertia of a rod of length a and mass m about an end is $ma^2/3$, so we must have

$$\frac{1}{2}\frac{ma^2}{3}\omega^2 = \frac{mga}{2}$$

which gives

▶ $$\omega = \sqrt{\dfrac{3g}{a}}.$$

When the rod breaks, we are told that no impulsive forces act so that the angular velocity of the upper part immediately after the break occurs is still $\sqrt{(3g/a)}$. However, the mass and length of this part of the rod are both half of the corresponding values for the unbroken rod, so the moment of inertia is one eighth of its former value, or $ma^2/24$. The kinetic energy of this part of the rod is thus

$$\frac{1}{2}\frac{ma^2}{24}\frac{3g}{a} = \frac{1}{16}mga.$$

If the rod now rotates through an angle θ, as shown in figure 33, its centre of mass will rise through a distance

$$\frac{a}{4}(1 - \cos\theta),$$

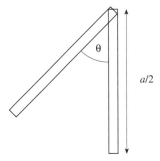

Figure 33

so it will gain potential energy

$$\frac{mga}{8}(1 - \cos\theta).$$

The largest value of θ reached by this part of the rod will occur when all the kinetic energy has been converted to potential energy. Thus

$$\frac{1}{16}mga = \frac{mga}{8}(1 - \cos\theta),$$

$$1 - \cos\theta = \frac{1}{2},$$

▶   hence $\theta = 60°$.

At the instant the rod breaks, shown in figure 34, the velocity of the upper end of its lower half is $\omega a/2$ to the left, and the velocity of the lower end is ωa, also to the left. The motion of this part at this instant can thus be resolved into a linear velocity $3\omega a/4$ to the left, and an angular velocity ω (clockwise) about its centre of mass. The centre of mass of the free fragment will thus follow a parabolic path downwards and to the left, while the fragment rotates clockwise at a constant angular velocity.

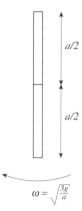

Figure 34

Problem 32

A uniform rod of length l and mass $2m$ rests on a smooth horizontal table. A point mass m moving horizontally at right angles to the rod with an initial velocity V collides with one end of the rod and sticks to it. Determine (a) the angular velocity of the system after the collision, (b) the position of the point on the rod which remains stationary immediately after the collision, and (c) the change in kinetic energy of the system as a whole as a result of the collision.

Solution

Let us put u for the velocity of the system's centre of mass after the collision, and ω for its angular velocity about the centre of mass, as shown in figure 35.

The first thing we need to do is to find the position of the centre of mass of the composite system. If we let this be a distance x_0 from the end of the rod to which the mass sticks, we can find x_0 as follows:

We will use x to measure distance along the rod, with $x = 0$

Indicate the orientation of the plumb line in each case. You may assume the Earth to be a sphere of radius 6.4×10^3 km.

Solution

The problem is concerned with the motion of a body relative to a rotating frame of reference (the Earth), and such problems are often most easily dealt with by ignoring the rotation of the frame. This requires the introduction of two fictitious forces, namely the *centrifugal force*

$$m\omega \times (\mathbf{r} \times \omega)$$

and the *Coriolis force*

$$2m\mathbf{v} \times \omega.$$

In these expressions, m is the mass of the body, $\mathbf{v}$ is its velocity with respect to the rotating frame, and $\mathbf{r}$ is its position vector relative to the rotation axis. ω is the angular velocity of the rotating frame of reference.

(a) Figure 37 shows the aircraft, seen from above, when it is at the North Pole.

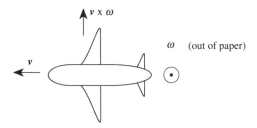

Figure 37

At the North Pole, the aircraft is flying perpendicular to the Earth's rotation axis (which is directed upwards through the North Pole), and the vector $\mathbf{r}$ is zero. The term $\mathbf{r} \times \omega$ is zero, so the centrifugal force is zero. The term $\mathbf{v} \times \omega$ has magnitude $v\omega$ and direction to the right of the aircraft's motion, so the plumb line will be deflected to the right. To calculate the deflexion, we can draw a view from behind to show the forces acting on the plumb bob, as shown in figure 38.

Balancing horizontal components gives

$$T \sin \theta = 2mv\omega$$

and balancing vertical components gives

$$T \cos \theta = mg,$$

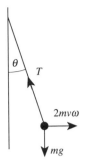

Figure 38

so $\tan \theta = 2v\omega/g$. Taking $v = 900$ km h$^{-1} = 250$ m s^{-1},
$\omega = 2\pi/(24 \times 60 \times 60)$ s^{-1} [see note below] $= 7.272 \times 10^{-5}$ s^{-1} and
$g = 9.81$ m s^{-2} gives $\tan \theta = 3.7 \times 10^{-3}$. Thus the deflexion angle θ is 3.7
▶ milliradians $= 0.21$ degrees. The direction of the deflexion is to the right
of the aircraft's motion.

(b) As the aircraft crosses the equator flying south, the velocity vector
v is directed oppositely to the angular velocity vector **ω**, and the position
vector **r** points upwards with magnitude R (the Earth's radius). Again, it
is useful to draw the view from above, shown in figure 39.

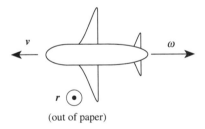

(out of paper)

Figure 39

The Coriolis force is clearly zero, and the centrifugal force acts
vertically upwards. Since there is no sideways force acting upon the
▶ plumb bob, the plumb line has a deflexion of zero.

(c) When the aircraft crosses latitude 45 °N (still flying southwards), it
is easiest to visualise the fictitious forces relative to the Earth rather than
relative to the aircraft, as shown in figure 40.

The vector **v** $\times$ **ω** is directed into the paper and has magnitude $v\omega/\sqrt{2}$,
so the Coriolis force on the plumb bob has magnitude $mv\omega\sqrt{2}$ and
direction to the right relative to the aircraft's motion. The vector **r** $\times$ **ω** is
directed into the paper and has magnitude $R\omega/\sqrt{2}$, so the vector

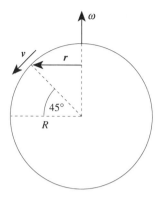

Figure 40

$\boldsymbol{\omega} \times (\mathbf{r} \times \boldsymbol{\omega})$ is directed to the left (on the diagram), with magnitude $R\omega^2/\sqrt{2}$. Relative to the aircraft's motion, the centrifugal force is thus of magnitude $mR\omega^2/\sqrt{2}$ in a direction 45° above the forward horizontal. This can be resolved into a component $mR\omega^2/2$ forwards, and a component $mR\omega^2/2$ upwards. The plumb line will thus be deflected forwards and to the right (i.e. approximately south-west).

We can calculate the deflexion in the same way as we did in part (a). The effect of the upward component of the centrifugal force is to reduce the apparent value of the acceleration due to gravity by an amount $R\omega^2/2 = 0.02 \, \mathrm{m\,s^{-2}}$, so we can allow for this by taking $g' = 9.79 \, \mathrm{m\,s^{-2}}$. The deflexion to the right is thus given by

$$\tan\theta = \frac{v\omega\sqrt{2}}{g'},$$

▶ which gives $\tan\theta = 0.0026$, so the deflexion to the right is 2.6 milliradians or 0.15 degrees. The deflexion forwards is given by

$$\tan\theta = \frac{R\omega^2}{2g'},$$

▶ which gives $\tan\theta = 0.0017$, so the forward deflexion is 1.7 milliradians or 0.10 degrees. These can be combined to calculate the total deflexion as $(0.15^2 + 0.10^2)^{1/2}$ degrees = 0.18 degrees, in a direction $\tan^{-1}(0.15/0.10) \approx 56$ degrees west of south.

[In fact, our assumption that the Earth's angular velocity ω is 2π radians in 24 hours is not strictly accurate. The Earth takes 24 hours to rotate once with respect to the frame which rotates round the sun once per year. Relative to an inertial frame, therefore, the Earth is rotating faster, by about one part in 365. The Earth's rotation period is thus approximately 23.93 days, which is called a *sidereal day*. The difference of 0.3% between the sidereal and solar days will not make a significant difference to our calculations.]

Problem 34

Find an expression for the deviation in angle between a stationary plumbline and the local vertical at a latitude ϕ, and calculate the maximum value of this deflexion. Assume the Earth to be a uniform sphere of radius 6400 km.

Solution

This is simpler than the previous problem since the plumb bob is stationary with respect to the Earth's surface, so the Coriolis force is zero. The forces acting on the bob are thus its weight mg, the centrifugal force $m\omega^2 R \cos \phi$, and the tension T in the string, as shown in figure 41.

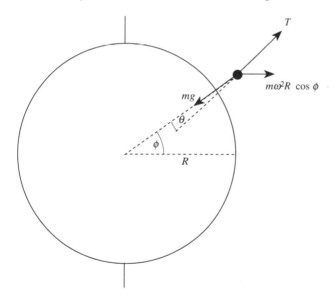

Figure 41

Resolving parallel to the local vertical,

$$T \cos \theta + m\omega^2 R \cos^2 \phi = mg,$$

and resolving horizontally,

$$T \sin \theta = m\omega^2 R \cos \phi \sin \phi.$$

Eliminating T and m gives

$$\cot \theta = \frac{g - \omega^2 R \cos^2 \phi}{\omega^2 R \cos \phi \sin \phi},$$

and since $g \gg \omega^2 R$, we can ignore the $\cos^2 \phi$ term and write $\cot \theta = 1/\tan \theta \approx 1/\theta$ to give

▶
$$\theta \approx \frac{\omega^2 R \cos \phi \sin \phi}{g} = \frac{\omega^2 R}{2g} \sin 2\phi.$$

This angular deflexion is southwards in the northern hemisphere and conversely, and has a maximum value at latitude $45°$ N or S. This maximum value is clearly

$$\frac{\omega^2 R}{2g}.$$

Taking $\omega = 7.292 \times 10^{-5}\,\mathrm{s}^{-1}$ (one revolution per sidereal day), $g = 9.81\,\mathrm{m\,s}^{-2}$ and $R = 6.4 \times 10^6\,\mathrm{m}$ as before gives a maximum deflexion

▶ of 1.7 milliradians (0.099 degrees).

[If we do not want to make the approximation $g \gg \omega^2 R$, we can differentiate our exact expression for $\cot \theta$ to obtain (after a little algebra)

$$\frac{d \cot \theta}{d\phi} = 2 - \frac{(g/\omega^2 R - \cos^2 \phi)(\cos^2 \phi - \sin^2 \phi)}{\cos^2 \phi \sin^2 \phi}.$$

Setting this equal to zero to find the minimum value of $\cot \theta$ (and hence the maximum value of θ) gives (after a little more algebra)

$$\cos^2 \phi = \frac{g}{2g - \omega^2 R}.$$

Substituting the values for ω, R and g gives $\phi = 44.95°$ and hence $\theta = 0.100$ degrees. We can see that the approximation was well justified.]

Problem 35

Derive the relationship between the impact parameter and the scattering angle for Rutherford scattering.

In a Rutherford scattering experiment, 4 MeV alpha-particles are incident on $^{197}_{79}\mathrm{Au}$ foil. Calculate the impact parameter which would give a deflexion of $10°$.

Solution

Figure 42 shows the geometry of Rutherford scattering.

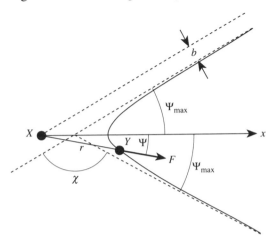

Figure 42

The particle Y enters at the top right travelling along a path which, if undeflected, would miss the deflecting particle X by a distance b (the impact parameter). We can describe the position of the particle Y at an arbitrary time using polar coordinates (r, ψ), and for simplicity we will write

$$F = \frac{B}{r^2}$$

for the repulsive force exerted on Y.

Let us introduce an x-axis along the line of symmetry. If the mass of the particle is m, we have

$$m\frac{d^2x}{dt^2} = \frac{B}{r^2}\cos\psi = \frac{B}{r^2\dfrac{d\psi}{dt}}\cos\psi\frac{d\psi}{dt}.$$

Now $mr^2 d\psi/dt$ is equal to the angular momentum J of the particle Y about the deflecting particle X, and it is a constant, so we can write

$$\frac{d^2x}{dt^2} = \frac{B}{J}\cos\psi\frac{d\psi}{dt}.$$

Integrating this with respect to time gives

$$\frac{dx}{dt} = \frac{B}{J}\sin\psi$$

(the constant of integration is zero because $dx/dt = 0$ when $\psi = 0$, by symmetry).

A long time after the particle Y has been deflected by the particle X, it will be travelling in a constant direction ψ_{max} and dx/dt will therefore have a value of $v_\infty \cos \psi_{max}$, where v_∞ is the velocity at infinite distance from X. Thus we have

$$v_\infty \cos \psi_{max} = \frac{B}{J} \sin \psi_{max},$$

so

$$\tan \psi_{max} = \frac{J v_\infty}{B}.$$

However, we also know that the angular momentum J of the particle Y is given by $m v_\infty b$ where b is the impact parameter, so we can write

$$\tan \psi_{max} = \frac{m v_\infty^2 b}{B}.$$

Finally, we note that the scattering angle χ is given by

$$2\psi_{max} + \chi = \pi,$$

so $\psi_{max} = \pi/2 - \chi/2$. Substituting for χ thus gives

$$\blacktriangleright \qquad \cot \frac{\chi}{2} = \frac{m v_\infty^2 b}{B}.$$

The charge on an alpha-particle is $+2e$ and the charge on an atomic nucleus of atomic number Z is $+Ze$, so the repulsive force between them at a distance r is

$$\frac{2 Z e^2}{4\pi \varepsilon_0 r^2}.$$

The parameter B is thus given by

$$B = \frac{Z e^2}{2\pi \varepsilon_0},$$

and the impact parameter b required to give a deflexion χ is

$$\frac{Z e^2}{4\pi \varepsilon_0 E} \cot \frac{\chi}{2},$$

where E is the initial kinetic energy of the alpha particle. Taking $Z = 79$, $E = 4 \times 10^6 \times 1.6 \times 10^{-19}$ J and $\chi = 10°$ gives $b = 3.2 \times 10^{-13}$ m.

Problem 36

An alpha-particle, of mass m and charge $+2e$, moves in the force field of a heavy nucleus of charge $+Ze$. It is initially at a large distance from the nucleus and travelling with speed v ($\ll c$) along a path which if continued without deviation would pass at a distance b from the nucleus. What is the distance of closest approach of the actual path of the particle?

Solution

As in the previous problem, let us write the electrostatic repulsion force as

$$F = \frac{B}{r^2},$$

where

$$B = \frac{2Ze^2}{4\pi\varepsilon_0}.$$

The electrostatic potential energy of the alpha-particle, when it is a distance r from the nucleus, is

$$\frac{B}{r}.$$

Let us write a for the closest distance of approach, and u for the speed of the alpha-particle when it is distance a from the nucleus. It is clear that the angular momentum of the alpha-particle about the nucleus is mvb when the particle is a large distance from the nucleus, and since no external forces act on the system this quantity must be conserved. When the particle is at its closest distance a from the nucleus it must be travelling perpendicular to its radius vector, so the angular momentum is given by mua. Thus, by conservation of angular momentum, we must have

$$u = \frac{vb}{a}.$$

At infinite distance from the nucleus, the potential energy of the alpha-particle is zero and its kinetic energy is $mv^2/2$. At its closest distance a, the kinetic energy is

$$\frac{1}{2}mu^2 = \frac{1}{2}m\left(\frac{vb}{a}\right)^2,$$

and the potentital energy is

$$\frac{B}{a},$$

so by conservation of energy we have

$$\frac{1}{2}mv^2 = \frac{1}{2}m\left(\frac{vb}{a}\right)^2 + \frac{B}{a}.$$

This can be rewritten as a quadratic in a:

$$a^2 - \frac{2B}{mv^2}a - b^2 = 0,$$

whose solutions are

$$a = \frac{B}{mv^2} \pm \sqrt{\left(\left(\frac{B}{mv^2}\right)^2 + b^2\right)}.$$

It is clear that only one of these solutions is positive and therefore meaningful, so if we select this solution and substitute into it our expression for B we finally obtain our solution

▶
$$a = \frac{Ze^2}{2\pi\varepsilon_0 mv^2} + \sqrt{\left(\frac{Z^2e^4}{4\pi^2\varepsilon_0^2m^2v^4} + b^2\right)}.$$

We can see that this is reasonable, because if we set $Z = 0$ (no electrostatic repulsion) we obtain $a = b$, and if we set $b = 0$ (head-on collision) we obtain $a = Ze^2/\pi\varepsilon_0 mv^2$, both of which results are easily verified by a simpler calculation.

Problem 37

A cylindrical rod has length L, radius r and shear modulus n. One end of the rod is clamped. Show that a torque C applied to the other end will twist it through an angle ϕ, where

$$\phi = \frac{2LC}{\pi n r^4}.$$

An engine is transmitting a power of 75 kW at 1100 r.p.m. through a cylindrical drive shaft of radius 0.025 m and length 2.0 m. If the shear modulus of the material of the drive shaft is 80 GPa, calculate the angle through which it is twisted.

Solution

Consider a cylindrical shell of radius a and thickness da, as shown in figure 43.

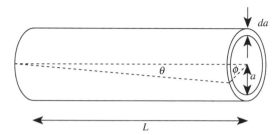

Figure 43

The angle of shear θ is given by $\phi a/L$, so the shear stress is equal to $n\phi a/L$. It is thus a function of position within the rod. Let us now consider the cross-section of this element, shown in figure 44.

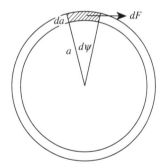

Figure 44

The element $d\psi$ has an area $a\,da\,d\psi$, so the force acting on it is

$$dF = \frac{n\phi a}{L}a\,da\,d\psi.$$

The contribution to the torque is $dC = a\,dF$, so

$$dC = \frac{n\phi}{L}a^3 da\,d\psi.$$

We can now integrate this with respect to a (from 0 to r) and ψ (from 0 to 2π) to obtain the total torque:

$$C = \frac{n\phi}{L}\int_0^r a^3\,da \int_0^{2\pi} d\psi = \frac{n\phi}{L}\frac{r^4}{4}2\pi = \frac{\pi nr^4\phi}{2L}.$$

Rearranging to find ϕ in terms of C gives

▶
$$\phi = \frac{2LC}{\pi n r^4}$$

as required.

▶   The drive shaft is rotating at 1100 r.p.m., so its angular velocity ω is $2\pi(1100/60)$ s^{-1} = 115 s^{-1}. Since the power being transmitted is 75 kW, the torque on the rod must be $(7.5 \times 10^4)/115$ N m = 652 N m. Substituting $C = 652$ N m, $L = 2.0$ m, $n = 8 \times 10^{10}$ N m^{-2} and $r = 0.025$ m into our expression for ϕ gives 0.027 radians (1.5 degrees).

Problem 38

Estimate the pressure at the bottom of the Marianas Trench (11.0 km below sea level). How large an error would be caused by neglecting the compressibility of sea water? Assume that the density of sea water is 1025 kg m^{-3} at sea level and that its bulk modulus is 2.1 GPa.

Solution

The variation of pressure with depth in a stationary fluid is given by

$$\frac{dp}{dz} = \rho g,$$

where p is the pressure, z is the depth, ρ is the density and g is the acceleration due to gravity. The bulk modulus B is defined as

$$B = \frac{-dp}{dV/V},$$

where dV/V is the fractional change in volume of a sample subjected to an isotropic pressure increase dp. If we consider a sample of the fluid having mass M, its volume $V = M/\rho$ so that

$$dV = -\frac{M}{\rho^2}d\rho$$

and hence $dV/V = -d\rho/\rho$. Combining these equations to eliminate dp gives

$$\frac{Bd\rho}{\rho} = \rho g \, dz.$$

This differential equation can be solved by rearranging and integrating it:

$$\int_{\rho_0}^{\rho} \frac{d\rho}{\rho^2} = \int_0^z \frac{g\, dz}{B},$$

where ρ_0 is the density at $z = 0$ and ρ is the density at depth z. Evaluating the integrals gives the following expression relating the density to the depth:

$$\frac{1}{\rho_0} - \frac{1}{\rho} = \frac{gz}{B}. \tag{1}$$

We can now use this expression to find the variation of pressure with depth. Since

$$dp = \frac{B\, d\rho}{\rho},$$

the relationship between p and ρ must be

$$p - p_0 = B \ln \frac{\rho}{\rho_0}, \tag{2}$$

where p_0 is the pressure at $z = 0$. If we multiply equation (1) by ρ_0 we obtain

$$1 - \frac{\rho_0}{\rho} = \frac{\rho_0 gz}{B},$$

so that

$$\ln \frac{\rho}{\rho_0} = -\ln\left(1 - \frac{\rho_0 gz}{B}\right).$$

Substituting this into equation (2) gives

$$p = p_0 - B \ln\left(1 - \frac{\rho_0 gz}{B}\right).$$

Inserting the values $p_0 = 1.01 \times 10^5$ Pa, $B = 2.1 \times 10^9$ Pa, $\rho_0 = 1.025 \times 10^3$ kg m^{-3}, $g = 9.81$ m s^{-2} and $z = 1.10 \times 10^4$ m, we obtain $p = 1.14 \times 10^8$ Pa (1.12×10^3 atmospheres). If we had ignored the compressibility of water the pressure at depth z would have been $p_0 + \rho_0 gz = 1.11 \times 10^8$ Pa, so we would have underestimated the pressure by 3×10^6 Pa (about 3%).

Problem 39

Show that for a monatomic model solid the Young modulus E is given approximately by

$$E = \frac{1}{r_0}\left(\frac{d^2U}{dr^2}\right)_{r=r_0},$$

where U is the interaction potential energy of a pair of atoms distance r apart and r_0 is their equilibrium separation.

When a uniform beam is bent, the bending moment M is given by

$$M = \frac{EI}{R},$$

where R is the radius of curvature of the arc and I is the second moment of area of the cross-section about the neutral axis. Use this relation to show that when a compressive stress is applied to the ends of a straight beam it will buckle when the stress reaches a critical value. Obtain an expression for this critical stress.

A uniform round rod of length 1 m and radius 10 mm is subjected to a compressive stress. At what value of the compressive *strain* will the rod begin to buckle? (I for a uniform circular disc of radius R about its diameter is given by $\pi R^4/4$.)

Solution

The bonds between the atoms of a solid can be modelled as springs of spring constant k. The shape of the interatomic potential function $U(r)$ determines the value of k as follows (see problem 19): The force F between a pair of atoms is given by

$$F = -\frac{dU}{dr},$$

so dF/dr is $-d^2U/dr^2$. If we evaluate this at the equilibrium separation r_0 and recognise that a restoring force (positive k) corresponds to negative F (since the force is directed opposite to the displacement), we can see that

$$k = \left(\frac{d^2U}{dr^2}\right)_{r=r_0}.$$

Now let us consider a macroscopic specimen of the solid of length L and cross-sectional area A. If the atoms have a simple cubic arrangement, the specimen has a length of L/r_0 bonds. Stretching the specimen by dx will increase the length of each of these bonds by $r_0 dx/L$, so the tension in each chain of atoms will be $kr_0 dx/L$. However, the specimen will contain A/r_0^2 of these chains of atoms, so the total force required to produce an extension dx must be $kA dx/Lr_0$. The stress in the specimen is thus kdx/Lr_0, and dividing this by the strain dx/L we see that the Young modulus is given by k/r_0. Thus we have

▶
$$E = \frac{1}{r_0}\left[\frac{d^2U}{dr^2}\right]_{r=r_0}$$

as required.

Let us assume that the beam has length l, and that the compressive forces F have succeeded in bending the beam slightly (very much exaggerated in figure 45), such that its midpoint has been displaced through a distance a ($\ll l$). The bending moment M at a point on the beam is clearly given by

$$M = F(a - y),$$

so the curvature is given by

$$\frac{1}{R} = \frac{F}{EI}(a - y).$$

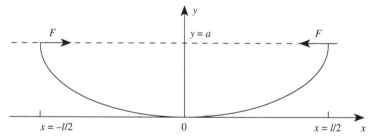

Figure 45

Now as long as $|dy/dx|$ is much less than unity, $1/R \approx d^2y/dx^2$, so the configuration of the beam satisfies the following differential equation:

$$\frac{d^2y}{dx^2} = \frac{F}{EI}(a - y).$$

This is somewhat easier to solve if we make the substitution $\eta = a - y$, giving

$$\frac{d^2\eta}{dx^2} = -\frac{F}{EI}\eta,$$

and we recognise that the general solution of this equation is

$$\eta = A \sin(kx) + B \cos(kx), \quad \text{where } k^2 = \frac{F}{EI}.$$

To find the values of A and B we consider the boundary conditions at $x = 0$ and $x = \pm l/2$.

At $x = 0$, $y = 0$ so $\eta = a$ giving $B = a$.

At $x = 0$, $dy/dx = 0$ so $d\eta/dx = 0$ giving $A = 0$.

Thus $\eta = a \cos(kx)$, but at $x = \pm l/2$ we know that $y = a$, so $\eta = 0$, and this requires that $kl/2 = \pi/2$. Since k depends on F, we have found the value of F we were looking for:

$$F = k^2 EI = \frac{\pi^2 EI}{l^2}.$$

[This is the force at which the beam will begin to buckle. The effect of increasing the force is to increase the value of a such that the simplifying assumptions of our analysis are no longer valid.]

If the cross-sectional area of the beam is a, the critical stress is

▶
$$\frac{\pi^2 EI}{al^2}.$$

The critical strain is therefore

$$\frac{\pi^2 I}{al^2},$$

so if the beam has the form of a rod of radius R, we can put $I = \pi R^4/4$ and $a = \pi R^2$ to obtain the expression $\pi^2 R^2/4l^2$ for the critical strain.

▶ Taking $l = 1$ m and $R = 10$ mm gives a value of 2.5×10^{-4} for this strain.

Problem 40

Two particles of mass m are connected by a light inextensible string of length l. One of the particles rests on a smooth horizontal table in which there is a small hole. The string passes through the hole so that the second particle hangs vertically below the hole. Take the position of the hole as the origin and describe the position of the particle on the table by plane polar coordinates r and θ, as shown in figure 46.

(i) Write down a formula for the angular momentum of the particle on the table about the origin in terms of r and $d\theta/dt$.

(ii) Explain why this angular momentum is a constant.

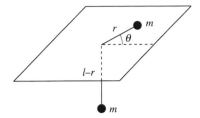

Figure 46

(iii) Write down a formula for the sum of the energies of both masses in terms of r, dr/dt and $d\theta/dt$. Using the result that this energy is constant, eliminate $d\theta/dt$ and show that

$$\left(\frac{dr}{dt}\right)^2 = \gamma - \frac{\delta}{r^2} - gr,$$

where γ and δ are constants and g is the acceleration due to gravity.

(iv) Initially the particle on the table is at a distance $l/2$ from the origin and is travelling with speed α in a direction at right angles to the string. Obtain a formula for $(dr/dt)^2$ when $r = l$.

(v) Hence find the condition such that the particle initially below the table does not pass through the hole. (The string always remains taut.)

Solution

▶   (i) The tangential speed of the mass on the table is $r(d\theta/dt)$ so its angular momentum J is $mr^2(d\theta/dt)$.

▶   (ii) No torque acts on the system so J must be constant.

(iii) The kinetic energy of the mass on the table is

$$\frac{1}{2}m\left[\left(\frac{dr}{dt}\right)^2 + r^2\left(\frac{d\theta}{dt}\right)^2\right]$$

and its gravitational potential energy is constant. The speed of the suspended mass is dr/dt so its kinetic energy is

$$\frac{1}{2}m\left(\frac{dr}{dt}\right)^2.$$

Since it hangs a distance $l - r$ below the table we can write its potential energy as

$$\text{constant} - mg(l - r),$$

which is equal to a different constant $+mgr$. Ignoring constant additive terms, we can thus write the total energy E of the system as

▶
$$E = m\left[\left(\frac{dr}{dt}\right)^2 + gr + \frac{r^2}{2}\left(\frac{d\theta}{dt}\right)^2\right].$$

Substituting $J = mr^2(d\theta/dt)$ to eliminate $d\theta/dt$, this can be rewritten as

$$\frac{E}{m} = \left(\frac{dr}{dt}\right)^2 + gr + \frac{J^2}{2m^2r^2} = \gamma,$$

where γ is a constant. Rearranging gives

▶
$$\left(\frac{dr}{dt}\right)^2 = \gamma - \frac{\delta}{r^2} - gr$$

as required, where $\delta = J^2/2m^2$ which is also a constant.

(iv) The initial conditions are

$$r = l/2,$$
$$(dr/dt) = 0 \text{ and}$$
$$r(d\theta/dt) = \alpha$$

so that

$$d\theta/dt = 2\alpha/l.$$

The angular momentum J is thus equal to $ml\alpha/2$, so the constant δ is equal to $l^2\alpha^2/8$. We can thus evaluate γ:

$$\gamma = \left(\frac{dr}{dt}\right)^2 + \frac{\delta}{r^2} + gr = 0 + \frac{l^2\alpha^2}{8}\cdot\frac{4}{l^2} + \frac{gl}{2} = \frac{\alpha^2}{2} + \frac{gl}{2}.$$

The value of $(dr/dt)^2$ when $r = l$ is then given by

▶
$$\gamma - \delta/l^2 - gl = \alpha^2/2 + gl/2 - \alpha^2/8 - gl = 3\alpha^2/8 - gl/2.$$

(v) The condition that the lower particle does not pass through the hole is clearly equivalent to the condition that dr/dt falls to zero for $r < l$. The limiting case is therefore $3\alpha^2/8 = gl/2$, and the condition we require is

▶
$$\alpha^2 < \frac{4gl}{3}.$$

Problem 41

A particle of mass m moves in a central force field such that its potential energy is given by $V = kr^n$, where r is its distance from the centre of force and k and n are constants. Find the conditions for a stable circular orbit.

Solution

The motion of the particle must be confined to a plane, so the problem is a two-dimensional one. Clearly the most useful system of coordinates in which to describe the motion will be polar coordinates, so let us begin by writing down an expression for the kinetic energy:

$$E_k = \frac{1}{2}m\left(\frac{dr}{dt}\right)^2 + \frac{1}{2}mr^2\left(\frac{d\theta}{dt}\right)^2.$$

Since the particle's potential energy is $V = kr^n$, its total energy is given by

$$E = \frac{1}{2}m\left(\frac{dr}{dt}\right)^2 + \frac{1}{2}mr^2\left(\frac{d\theta}{dt}\right)^2 + kr^n,$$

and as no external forces act on the system, this quantity is conserved. The angular momentum J is also conserved, and since J is given by $mr^2(d\theta/dt)$ we can use this fact to rewrite $(d\theta/dt)$ in terms of r:

$$E = \frac{1}{2}m\left(\frac{dr}{dt}\right)^2 + \frac{J^2}{2mr^2} + kr^n.$$

This expression looks like (has the mathematical form of) the total energy of a particle of mass m moving in a *single* dimension such that its effective potential energy V' is

$$V' = \frac{J^2}{2mr^2} + kr^n.$$

A circular orbit corresponds to $r = r_0 = $ constant, in other words to a position of equilibrium in the one-dimensional problem. We know that this equilibrium position is the value of r for which $dV'/dr = 0$, so let us evaluate this differential coefficient:

$$\frac{dV'}{dr} = -\frac{J^2}{mr^3} + nkr^{n-1}.$$

Setting this equal to 0 at $r = r_0$ gives

$$r_0^{n+2} = \frac{J^2}{mnk}.$$

Since r_0, J and m are all positive, this gives us our first condition, which is that n and k must have the same sign. [We could have written this down earlier, since it is just the condition that the central force is attractive.]

For the equilibrium to be stable, we must have $d^2V'/dr^2 > 0$ at $r = r_0$. Differentiating again gives

$$\frac{d^2V'}{dr^2} = \frac{3J^2}{mr^4} + n(n-1)kr^{n-2}.$$

Putting $r = r_0$ and $d^2V'/dr^2 > 0$, and multiplying through by r_0^4, gives

$$\frac{3J^2}{m} + n(n-1)kr_0^{n+2} > 0,$$

and substituting our expression for r_0^{n+2} gives

$$\frac{J^2}{m}(n+2) > 0.$$

The condition for stability is thus $n > -2$. Putting these results together, we have

▶ $n > 0,\ k > 0$

or

▶ $-2 < n < 0,\ k < 0$

as the conditions for stable circular orbits.

Problem 42

Give Lagrange's equations of motion. Applying them to the motion of a planet orbiting the Sun, give expressions for the kinetic and potential energies of the planet in polar coordinates and obtain two equations of motion for the radial and angular motion. Show that the angular equation of motion can be integrated and leads to the conservation of angular momentum.

By changing the radial coordinate r to $u = 1/r$ and eliminating time, show that the radial equation of motion has the form of a differential equation for displaced simple harmonic motion. Hence obtain a solution for the shape of the orbit.

Solution

Lagrange's equations of motion can be expressed as

$$\frac{d}{dt}\left(\frac{\partial L}{\partial p_i}\right) = \frac{\partial L}{\partial q_i},$$

where q_i is a position-like variable, $p_i = dq_i/dt$ is a momentum-like variable and L is the Lagrangian, defined as $T - V$ where T is the kinetic energy of the system and V is its potential energy.

For a planet of mass m in orbit about the Sun of mass M, the kinetic energy is given in polar coordinates by

▶ $$T = \frac{1}{2}m\left(\frac{dr}{dt}\right)^2 + \frac{1}{2}mr^2\left(\frac{d\theta}{dt}\right)^2$$

and the potential energy is given by

▶ $$V = -\frac{GMm}{r}.$$

We will thus take

$$q_1 = r,$$
$$p_1 = dr/dt,$$
$$q_2 = \theta,$$
$$p_2 = d\theta/dt$$

as the variables that describe the motion. The Lagrangian is thus

$$L = \frac{1}{2}mp_1^2 + \frac{1}{2}mq_1^2p_2^2 + \frac{GMm}{q_1}.$$

Considering first the radial variable (subscript 1),

$$\partial L/\partial p_1 = mp_1 = m\,dr/dt,$$

and

$$\partial L/\partial q_1 = mq_1p_2^2 - GMm/q_1^2 = mr(d\theta/dt)^2 - GMm/r^2.$$

Applying Lagrange's equation gives

▶ $$m\frac{d^2r}{dt^2} = mr\left(\frac{d\theta}{dt}\right)^2 - \frac{GMm}{r^2}. \tag{1}$$

Now considering the angular variable (subscript 2),

$$\partial L/\partial p_2 = mq_1^2p_2 = mr^2(d\theta/dt),$$

and

$$\partial L/\partial q_2 = 0.$$

Thus

$$\frac{d}{dt}\left(mr^2\frac{d\theta}{dt}\right) = 0.$$

This shows that

$$mr^2\frac{d\theta}{dt} = J \tag{2}$$

is a constant of the motion, and in fact we recognise it as the angular momentum of the planet about the Sun.

We want to use equations (1) and (2) to derive an expression for the shape of the planet's path in space, i.e. an expression for r as a function of θ with the time-dependencies of both r and θ cancelled out. The problem asks us to use the variable $u = 1/r$ instead of r itself. Thus

$$\frac{du}{d\theta} = \frac{du}{dr}\frac{dr}{dt}\frac{dt}{d\theta} = -\frac{1}{r^2}\frac{dr/dt}{d\theta/dt}.$$

Substituting from (2), we can replace $r^2(d\theta/dt)$ by the constant term J/m, so that

$$\frac{du}{d\theta} = -\frac{m}{J}\frac{dr}{dt}. \tag{3}$$

We can also calculate the value of $d^2u/d\theta^2$:

$$\frac{d^2u}{d\theta^2} = \frac{d}{d\theta}\left(\frac{du}{d\theta}\right) = \frac{dt}{d\theta}\frac{d}{dt}\left(\frac{du}{d\theta}\right).$$

Substituting from (3) gives

$$\frac{d^2u}{d\theta^2} = -\frac{m}{J}\frac{(d^2r/dt^2)}{(d\theta/dt)}. \tag{4}$$

Dividing (1) by $d\theta/dt$ gives

$$\frac{m(d^2r/dt^2)}{(d\theta/dt)} = mr\frac{d\theta}{dt} - \frac{GMm}{r^2(d\theta/dt)},$$

and substituting from (4) for the term on the left-hand side and from (2) for the terms on the right-hand side, this can be rewritten as

$$-J\frac{d^2u}{d\theta^2} = \frac{J}{r} - \frac{GMm^2}{J}.$$

Dividing through by $-J$ and recalling that $u = 1/r$, we obtain

$$\frac{d^2u}{d\theta^2} = -u + \frac{GMm^2}{J^2}. \tag{5}$$

This looks like the differential equation for simple harmonic motion, with the angle θ replacing the usual time variable, so we will try a solution of the form

$$u = A \cos(B\theta) + C,$$

where A, B and C are constants. This would give $d^2u/d\theta^2 = -AB^2 \cos(B\theta)$, so equation (5) is satisfied if $B = 1$ and $C = GMm^2/J^2$. The general solution of the equation of motion is thus

▶ $$u = \frac{1}{r} = A \cos\theta + C,$$

where A and C are constants. This is the polar equation of an ellipse.

[As we have shown, the constant $C = GMm^2/J^2$ is determined only by the angular momentum.]

Gravitation and orbits

Problem 43

In Arthur C. Clarke's novel *2001* an object in free space has the form of a very large plane slab 222 m thick. A test body released 100 m away from the surface of the slab, initially at rest relative to it, takes 20 minutes to fall to its surface. What is the density of the slab, assumed homogeneous?

Solution

If we ignore edge effects (as the problem hints that we should by telling us that the slab is 'very large'), it is clear that the gravitational field lines must be perpendicular to the slab and that the gravitational field strength g must be constant.

Gauss's theorem in gravitation states that

$$\int \mathbf{g} \cdot d\mathbf{s} = -4\pi G \sum M$$

where $\int \mathbf{g} \cdot d\mathbf{s}$ is the integral of the gravitational field $\mathbf{g}$ over a surface whose outward-pointing normal is $d\mathbf{s}$ and $\sum M$ is the mass enclosed within the surface. We can choose any Gaussian surface, but the most convenient is one which reflects the symmetry of the situation, so we will choose one which has sides perpendicular to the slab and ends of area A parallel to the slab, as shown in figure 47.

If the slab has thickness t and density ρ, the mass $\sum M$ enclosed by the Gaussian surface is $\rho t A$, and the contribution to $\int \mathbf{g} \cdot d\mathbf{s}$ is $-gA$ from each end and zero from the sides (since at the sides $d\mathbf{s}$ is perpendicular to $\mathbf{g}$). Applying Gauss's theorem thus gives

$$-2gA = -4\pi G\rho t A,$$

so

$$g = 2\pi G\rho t.$$

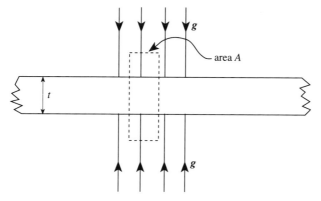

Figure 47

Now for a body falling from rest in a uniform gravitational field of strength g, the distance x fallen in time T is $gT^2/2$, thus

$$x = \frac{1}{2}2\pi G \rho t T^2,$$

which gives, on rearranging,

$$\rho = \frac{x}{\pi G t T^2}.$$

Substituting $x = 100\,\text{m}$, $t = 222\,\text{m}$ and $T = 20$ minutes $= 1200\,\text{s}$ gives
$\rho = 1.49 \times 10^3\,\text{kg}\,\text{m}^{-3}$.

Problem 44

A sphere of uniform density ρ has within it a spherical cavity whose centre is a distance a from the centre of the sphere. Show that the gravitational field within the cavity is uniform and determine its magnitude and direction.

Solution

By the principle of superposition, we know that the (vector) gravitational field in the cavity is equal to the gravitational field of the original sphere minus the gravitational field of the material which has been removed.

Let us consider first the gravitational field inside a sphere of density ρ.

At a radius r, the gravitational field is equal to the field due to the mass contained within radius r, so we can write

$$g = \frac{G\frac{4}{3}\pi r^3 \rho}{r^2} = \frac{4\pi G\rho r}{3},$$

and since the field is directed radially inwards, we can write this in vector form as

$$\mathbf{g} = -\frac{4}{3}\pi G\rho \mathbf{r}.$$

Now we can consider the sphere with a spherical cavity (see figure 48).

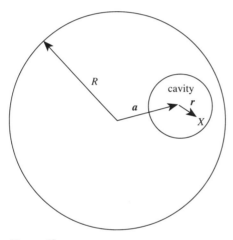

Figure 48

Consider a point X within the cavity such that the vector displacement of X from the centre of the cavity is $\mathbf{r}$. The vector displacement of X from the centre of the original sphere is thus $\mathbf{a} + \mathbf{r}$, and the gravitational field at X due to the original sphere is

$$-\frac{4}{3}\pi G\rho(\mathbf{a} + \mathbf{r}).$$

The gravitational field due to the material removed to make the cavity is

$$-\frac{4}{3}\pi G\rho \mathbf{r},$$

so the total gravitational field at X is

$$\mathbf{g} = -\frac{4}{3}\pi G\rho(\mathbf{r} + \mathbf{a}) + \frac{4}{3}\pi G\rho \mathbf{r} = -\frac{4}{3}\pi G\rho \mathbf{a}.$$

▶ The field within the cavity is thus uniform. Its magnitude is $4\pi G\rho a/3$ (i.e. depends only on the position, and not the size, of the cavity) and its direction is parallel to the line joining the centre of the cavity to the centre of the original sphere.

Problem 45

A uniform hollow sphere has internal radius a and external radius b. Taking the potential at infinity to be zero, show that the ratio of the gravitational potential at a point on the outer surface to that on the inner surface is

$$\frac{2(b^3 - a^3)}{3b(b^2 - a^2)}.$$

Solution

Figure 49 shows the sphere.

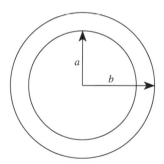

Figure 49

Suppose the density of the hollow sphere is ρ (this will cancel out later since we will be calculating a ratio). The mass M of the hollow sphere is clearly

$$M = \frac{4}{3}\pi\rho(b^3 - a^3)$$

and the potential at a point on the outer surface is $-GM/b$, so we may write

$$V(b) = -\frac{G\frac{4}{3}\pi\rho(b^3 - a^3)}{b}.$$

Next we consider a point at a distance r from the centre of the sphere, such that $a < r < b$. The magnitude of the gravitational field strength g at this point is that due to the mass contained within radius r, and since this mass is

$$\frac{4}{3}\pi\rho(r^3 - a^3)$$

the gravitational field strength is

$$g = -\frac{G\dfrac{4}{3}\pi\rho(r^3 - a^3)}{r^2},$$

where the negative sign shows that the field is directed radially inwards, i.e. opposite to the direction of increasing r.

Now we know that the relationship between g and the gravitational potential V is

$$g = -\frac{dV}{dr},$$

so

$$
\begin{aligned}
V(a) - V(b) &= -\int_b^a g\,dr = \frac{4\pi\rho G}{3}\int_b^a \frac{r^3 - a^3}{r^2}\,dr \\
&= \frac{4\pi\rho G}{3}\left[\frac{r^2}{2} + \frac{a^3}{r}\right]_b^a \\
&= \frac{4\pi\rho G}{3}\left(\frac{a^2}{2} - \frac{b^2}{2} + \frac{a^3}{a} - \frac{a^3}{b}\right) \\
&= \frac{4\pi\rho G}{3}\left(\frac{3a^2}{2} - \frac{b^2}{2} - \frac{a^3}{b}\right).
\end{aligned}
$$

Thus using our expression for $V(b)$ we obtain

$$
\begin{aligned}
V(a) &= \frac{4\pi\rho G}{3}\left(-b^2 + \frac{a^3}{b} + \frac{3a^2}{2} - \frac{b^2}{2} - \frac{a^3}{b}\right) \\
&= \frac{4\pi\rho G}{3}\left(-\frac{3b^2}{2} + \frac{3a^2}{2}\right),
\end{aligned}
$$

so that

$$
\frac{V(b)}{V(a)} = \frac{-\dfrac{4\pi\rho G}{3}\left(\dfrac{b^3 - a^3}{b}\right)}{\dfrac{4\pi\rho G}{3}\left(-\dfrac{3b^2}{2} + \dfrac{3a^2}{2}\right)} = \frac{2(b^3 - a^3)}{3b(b^2 - a^2)}
$$

as required.

Problem 46

A hole is bored in a straight line through the Earth from Cambridge to New York, and a ball-bearing is dropped in at the Cambridge end. Assuming that frictional and air resistance forces are negligible, and that the Earth may be taken as a uniform-density sphere of radius 6400 km, how long does it take the ball bearing to arrive in New York? [Neglect any effects due to the rotation of the Earth, and assume the acceleration due to gravity at the Earth's surface to be $9.8 \, \mathrm{m\,s^{-2}}$.]

Solution

The problem does not tell us the geographical locations of Cambridge and New York, and it is unlikely that we are expected to know them, so we should not be surprised if they do not figure in the answer.

Consider the situation when the ball-bearing is at the point P, a distance x from the mid-point of the tunnel (see figure 50). We will denote the perpendicular distance from the tunnel to the centre of the Earth (which we do not know) by a, and the radial distance of P from the Earth's centre by r.

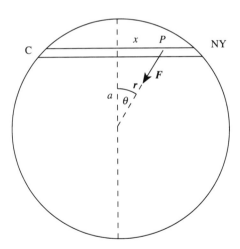

Figure 50

We know from Gauss's theorem that the gravitational field at a point P inside a uniform sphere is the same as the field which would be produced by all the matter of the sphere closer to the centre than P, so the force F acting on the ball-bearing is given by

$$F = \frac{\frac{4}{3}\pi r^3 \rho m G}{r^2},$$

where ρ is the Earth's density, m is the mass of the ball-bearing, and G is the gravitational constant. The component of this force acting parallel to the tunnel in the direction of increasing x is $-F \sin \theta$, so the acceleration of the ball-bearing is

$$\frac{d^2 x}{dt^2} = -\frac{4}{3}\pi \rho G r \sin \theta,$$

but the distance x is given by $r \sin \theta$, so we can write this as

$$\frac{d^2 x}{dt^2} = -\frac{4}{3}\pi \rho G x,$$

which is the equation of simple harmonic motion, and is (as we hoped) independent of a. The period of this motion is

$$2\pi \sqrt{\frac{3}{4 G \pi \rho}}$$

and the time T for the ball-bearing to get from Cambridge to New York will be half a cycle, so

$$T = \pi \sqrt{\frac{3}{4 G \pi \rho}}.$$

Now we can use the information we have been given about the Earth's radius and surface gravity to find the density ρ. If the radius is R, the surface gravitational acceleration must be

$$g = \frac{\frac{4}{3}\pi R^3 \rho G}{R^2} = \frac{4\pi R \rho G}{3},$$

so

$$\rho = \frac{3g}{4\pi R G}.$$

Substituting this into our expression for T gives

$$T = \pi \sqrt{\frac{R}{g}},$$

▶   so taking $R = 6.4 \times 10^6$ m and g $= 9.8 \, \mathrm{m\,s^{-2}}$ gives $T = 2.5 \times 10^3$ s
(about 42 minutes). [In fact this is half the orbital period of a satellite in a
low orbit round the Earth, ignoring the effects of air resistance.]

Problem 47

In 1910, on its sixth trip around the Sun after that of 1456, Halley's comet
was observed to pass near the Sun at a distance of 9.0×10^{10} m. Estimate
how far the comet travels from the Sun at the outer extreme of its orbit,
and determine the ratio of its maximum orbital speed to its minimum
speed.

Solution

The comet's period must be $(1910 - 1456)/6 = 75.7$ years, and we know
from Kepler's third law that the periods of bodies orbiting the Sun are
proportional to the 3/2 power of their semi-major axes, so the semi-major
axis of the comet's orbit must be $75.7^{2/3}$ times as great as the semi-major
axis of the Earth's orbit (since the Earth takes 1 year, by definition, to
orbit the Sun). Thus the semi-major axis a of the comet's orbit is 17.9
astronomical units, where the astronomical unit is defined as the
semi-major axis of the Earth's orbit.

We can calculate the length D of the astronomical unit as follows:
For an object of mass m in a circular orbit of radius D around the Sun
(mass M), the gravitational force GMm/D^2 can be equated to the
centripetal force $m\omega^2 D$, so that the angular velocity ω is given by

$$\omega^2 = \frac{GM}{D^3}.$$

Hence the period T is given by

$$T^2 = \frac{4\pi^2 D^3}{GM},$$

which can be rearranged to give

$$D^3 = \frac{GMT^2}{4\pi^2}.$$

We recall that this result is also true for an elliptical orbit if D is the
semi-major axis, so taking $T = 1$ year $= 3.16 \times 10^7$ s and $M = 2.0 \times 10^{30}$ kg gives $D = 1.50 \times 10^{11}$ m. The comet's semi-major axis a is thus
2.69×10^{12} m.

The comet's minimum distance r_{min} from the Sun is given by $a(1 - e)$ where e is the eccentricity (see figure 51), and has a value of 9.0×10^{10} m, so

$$e = 1 - \frac{9.0 \times 10^{10}}{2.69 \times 10^{12}} = 0.9665.$$

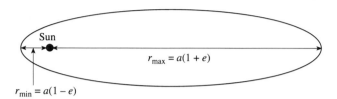

Figure 51

The maximum distance r_{max} from the Sun is given by $a(1 + e) = 2.69 \times 10^{12} \times 1.9665$ m $= 5.3 \times 10^{12}$ m.

[The problem did not ask us to calculate the eccentricity of the comet's orbit, and in fact we did not need to do so since r_{max} could have been obtained directly using $r_{max} = 2a - r_{min}$.]

The ratio of maximum to minimum orbital speed is equal to the ratio $r_{max}/r_{min} = (1 + e)/(1 - e) = 59$.

Problem 48

If gravitational forces alone prevent a spherical, rotating neutron star from disintegrating, estimate the minimum mean density of a star that has a rotation period of one second.

Solution

We will assume that the star remains spherical even when it is just about to disintegrate, and that the density is uniform (the former assumption is less reasonable than the latter). If the star has radius r and angular velocity ω, the centripetal acceleration of a particle at its equator is $\omega^2 r$. This must be provided by the gravitational force, otherwise the star will disintegrate, so we must have

$$\frac{G\frac{4}{3}\pi r^3 \rho}{r^2} > \omega^2 r.$$

Thus

$$\rho > \frac{3\omega^2}{4\pi G} = \frac{3\pi}{GT^2},$$

▶ where T is the star's rotation period. Taking $T = 1\,\text{s}$ gives $\rho > 1.4 \times 10^{11}\,\text{kg m}^{-3}$.

[In fact, the densities of neutron stars are about $5 \times 10^{17}\,\text{kg m}^{-3}$, so our simple formula suggests that they should have rotation periods of at least 0.5 milliseconds. The fastest observed pulsars (which are spinning neutron stars emitting beams of electromagnetic radiation) have periods of about 1 millisecond.]

Problem 49

An astronaut marooned on the surface of an asteroid, of radius r and mean density equal to that of the Earth, finds that he can escape by jumping. What is the maximum value of r?

Solution

We will assume that rotation effects can be ignored, and that the maximum height to which the astronaut could jump on the surface of the Earth is h. If the Earth's mass is M and its radius is R, the gravitational field strength at the surface is

$$g_E = \frac{GM}{R^2} = \frac{4\pi G\rho R}{3},$$

where ρ is the Earth's mean density. Thus the maximum kinetic energy which the astronaut can convert to potential energy in a jump is $mg_E h$, where m is his mass. We will assume that this quantity is the same for the astronaut on the asteroid.

At the surface of the asteroid, the astronaut's gravitational potential energy is

$$-\frac{GM'm}{r},$$

where M' is the mass of the asteroid and r is its radius. We can write this as

$$-\frac{4\pi Gmr^2\rho}{3},$$

where we have assumed that the asteroid's mean density is the same as the Earth's. Thus in order to be able to escape from the asteroid,

$$\frac{4\pi G\rho Rmh}{3} > \frac{4\pi Gmr^2\rho}{3} \cdot \quad \therefore r^2 < Rh;$$

i.e. the maximum value of r is $\sqrt{(Rh)}$. Taking $R = 6.4 \times 10^6$ m and
▶ guessing $h \approx 1$ m gives $r_{max} \approx 2.5$ km.

Problem 50

The planet Jupiter has a radius equal to 11.2 times the Earth's radius, a mass equal to 318 times the Earth's mass and a period of rotation about its axis of 10.2 hours. Calculate (a) the minimum velocity with which a rocket would need to leave the Jovian surface in order to escape entirely from the gravitational attraction of Jupiter, and (b) the radius of a circular orbit around Jupiter in which a satellite would remain above the same point on the Jovian surface. (You may take the escape velocity from the Earth to be 11.2 km s^{-1} and the radius of a geosynchronous orbit to be $44\,200$ km.)

Solution

(a) The gravitational potential energy of a body of mass m at the surface of a spherical planet of mass M and radius R is

$$-\frac{GMm}{R}.$$

[If the planet is rotating, the body will also have kinetic energy, but this term is negligible in comparison with the gravitational potential energy for both of the planets under discussion.] In order for the body to escape from the planet it must therefore be given an initial velocity v such that

$$\frac{1}{2}mv^2 = \frac{GMm}{R} \quad \therefore v \propto \sqrt{\frac{M}{R}}.$$

The escape velocity from Jupiter is therefore

▶ $$11.2 \times \sqrt{\frac{318}{11.2}} \text{ km s}^{-1} = 59.7 \text{ km s}^{-1}.$$

(b) A satellite of mass m in a circular orbit of radius r and angular velocity ω about the planet experiences a gravitational force

$$\frac{GMm}{r^2}$$

which provides the centripetal force $m\omega^2 r$. The angular velocity is thus given by

$$\omega^2 = \frac{GM}{r^3}.$$

If the orbit is geosynchronous and the planet has a rotation period of T, $\omega = 2\pi/T$. Inserting this relationship and rearranging gives

$$r^3 = \frac{GMT^2}{4\pi^2} \quad \therefore r \propto M^{1/3} T^{2/3}.$$

Since the Earth rotates once about its axis in 23.93 hours (see problem 33), the radius of a stationary orbit about Jupiter must be

$$\blacktriangleright \qquad 44.2 \times 10^3 \times 318^{1/3} \times \left(\frac{10.2}{23.93}\right)^{2/3} \text{ km} = 171 \times 10^3 \text{ km}.$$

Problem 51

A planet is in a circular orbit about a massive star. The star undergoes a spherically symmetric explosion in which one per cent of its mass is suddenly ejected to a distance well beyond that of the planet's orbit. Find the eccentricity of the new orbit of the planet, assuming the planet itself is unaffected by the explosion.

Solution

For definiteness, let us put M for the mass of the star before the explosion, r_0 for the radius of the planet's orbit, and m for the planet's mass. We can find the orbital speed v_0 by equating the gravitational force to the centripetal force:

$$\frac{GMm}{r_0^2} = \frac{mv_0^2}{r_0} \quad \therefore v_0^2 = \frac{GM}{r_0}.$$

Immediately after the explosion, the planet's position and velocity are unchanged (since to change the position in an infinitesimal time would need an infinite velocity, and to change the velocity would need an infinite acceleration), so we know the behaviour of the planet at one point in its new orbit. Since the planet is travelling perpendicularly to its radius vector at this point, this must be either the nearest or the furthest point of the new orbit from the star (it is actually the nearest). If we can find an expression for r_1, the distance from the star to the other point at which the planet is moving tangentially (see figure 52), we can find the eccentricity by comparing r_1 and r_0. The simplest way to do this is by considering the conservation of energy and of angular momentum.

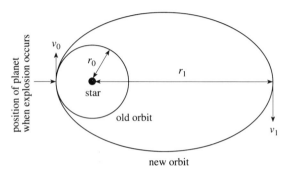

Figure 52

At distance r_0, the kinetic energy of the planet is $mv_0^2/2$ which we can write as

$$\frac{GMm}{2r_0}$$

using our expression for v_0. If we write M' for the new mass of the star, the potential energy of the planet at this position is

$$-\frac{GM'm}{r_0}.$$

Let us write v_1 for the speed of the planet at its furthest distance r_1 from the star. In order to conserve angular momentum we must have

$$mv_0r_0 = mv_1r_1 \quad \therefore v_1 = \frac{v_0r_0}{r_1}.$$

The planet's kinetic energy at r_1 is thus

$$\frac{1}{2}m\frac{v_0^2r_0^2}{r_1^2},$$

which we can write as

$$\frac{GMmr_0}{2r_1^2},$$

again using our expression for v_0. The potential energy at r_1 is

$$-\frac{GM'm}{r_1}.$$

Thus equating the total energy at r_0 with the total energy at r_1 we obtain

$$\frac{GMm}{2r_0} - \frac{GM'm}{r_0} = \frac{GMmr_0}{2r_1^2} - \frac{GM'm}{r_1}.$$

If we put μ for the dimensionless ratio M'/M and ρ for the dimensionless ratio r_1/r_0, this expression can be simplified to

$$1 - 2\mu = \frac{1}{\rho^2} - \frac{2\mu}{\rho}.$$

Multiplying through by ρ^2 gives

$$\rho^2(1 - 2\mu) + 2\mu\rho - 1 = 0,$$

which can be factorised as

$$(\rho - 1)(\rho[1 - 2\mu] + 1) = 0,$$

and since we can eliminate $\rho = 1$ as a solution, we must have

$$\rho = \frac{1}{2\mu - 1}.$$

[We can see from this, incidentally, that if $\mu < 1$, ρ must be greater than 1, so that r_1 is greater than r_0 as we stated earlier.]

Now the eccentricity e can be found from the expressions $r_0 = a(1 - e)$ and $r_1 = a(1 + e)$ where a is the semi-major axis of the ellipse. Taking the ratio of these expressions to eliminate a (which we do not want to know) we find that

$$\rho = \frac{1 + e}{1 - e}, \quad \text{so } e = \frac{\rho - 1}{\rho + 1}.$$

Substituting our expression for ρ gives

$$e = \frac{1 - \mu}{\mu},$$

▶   and taking $\mu = 0.99$ gives $e = 0.01$.

Special relativity

Problem 52

In a certain inertial frame two light pulses are emitted, a distance 5 km apart and separated by 5 μs. An observer who is travelling, parallel to the line joining the points where the pulses are emitted, at a velocity v with respect to this frame notes that the pulses are simultaneous. Find v.

Solution

The 'standard configuration' in special relativity problems involves two inertial frames S and S' such that, according to observers stationary with respect to the frame S, the frame S' has a velocity v in the x-direction. If $\Delta x, \Delta y, \Delta z, \Delta t$ are the intervals measured in S between two events, and $\Delta x', \Delta y', \Delta z', \Delta t'$ are the intervals between the same events measured in S', the relations between the intervals are given by the Lorentz transformations:

$$
\begin{aligned}
\Delta x' &= \gamma(\Delta x - v\Delta t), & \Delta x &= \gamma(\Delta x' + v\Delta t'), \\
\Delta y' &= \Delta y, & \Delta y &= \Delta y', \\
\Delta z' &= \Delta z, & \Delta z &= \Delta z', \\
\Delta t' &= \gamma(\Delta t - v\Delta x/c^2), & \Delta t &= \gamma(\Delta t' + v\Delta x'/c^2),
\end{aligned}
$$

where

$$
\gamma = \sqrt{\frac{1}{1 - v^2/c^2}}.
$$

We will assume that S is the frame in which the pulses are emitted with a time separation of 5 μs, so that $\Delta x = 5$ km, $\Delta t = 5\ \mu$s. We require to find the frame S' in which $\Delta t' = 0$. From the Lorentz transformation, we can see that this is so if

$$
v = c^2 \Delta t / \Delta x.
$$

▶ Inserting the values of Δt and Δx gives $v = 9 \times 10^7\ \mathrm{m\,s^{-1}}\ (= 0.3\,c)$.

Problem 53

Observer A sees two events at the same place ($\Delta x = \Delta y = \Delta z = 0$) and separated in time by $\Delta t = 10^{-6}$ s. A second observer B sees them to be separated by $\Delta t' = 2 \times 10^{-6}$ s. What is the separation in space of the two events according to B? What is the speed of B relative to A?

Solution

Observer A is at rest in frame S, and observer B is at rest in the frame S'. The Lorentz transformation for $\Delta t'$ gives

$$\Delta t' = \gamma \Delta t$$

(since $\Delta x = 0$), so we must have $\gamma = 2$. Now since

$$\gamma = \sqrt{\frac{1}{1 - v^2/c^2}}$$

it follows that

$$v = c\sqrt{\left(1 - \frac{1}{\gamma^2}\right)},$$

▶ so $v = c(1 - 1/4)^{1/2} = \sqrt{3}\,c/2$. The Lorentz transformation for $\Delta x'$ gives

$$\Delta x' = -\gamma v \Delta t$$

(again using the fact that $\Delta x = 0$), so

$$\Delta x' = -2\frac{\sqrt{3}}{2}3 \times 10^8 \times 10^{-6} \text{ m} = -520 \text{ m}.$$

▶ Thus according to observer B, the spatial separation of the two events is $\Delta x' = -520$ m, $\Delta y' = \Delta z' = 0$.
   [We could have calculated the magnitude of $\Delta x'$ directly, without first calculating v, by using the *Lorentz invariant interval*. This is defined as

$$\Delta s^2 = \Delta x^2 + \Delta y^2 + \Delta z^2 - c^2 \Delta t^2,$$

and it can be shown that $\Delta s^2 = \Delta s'^2$, i.e. that the interval between two events is the same in any inertial frame. In this problem, $\Delta y' = \Delta z' = 0$ so that

$$0 - c^2 \Delta t^2 = \Delta x'^2 - c^2 \Delta t'^2,$$

which can be rearranged to give

$$\Delta x'^2 = c^2(\Delta t'^2 - \Delta t^2).$$

Putting $\Delta t' = 2 \,\mu$s and $\Delta t = 1 \,\mu$s gives $\Delta x' = \pm 520$ m.]

Problem 54

Two inertial frames of reference S and S' are in the standard configuration, frame S' having velocity v with respect to frame S. At the instant when their spatial origins O and O' coincide, a light beam is emitted from O and O' along the positive x- and x'-axis. The beam is reflected by a mirror M fixed in S at a distance d from O and with its plane perpendicular to the x-axis. Consider the following three events:

    (1) light beam reaches M,
    (2) reflected beam returns to O',
    (3) reflected beam returns to O.

Calculate the times of these events as measured by observers in frame S. Use the Lorentz transformation to determine the times of these events as measured by observers in frame S'. Show how observers in frame S' would explain their measurements without reference to frame S.

Solution

It will be helpful to draw a space–time diagram of the events in S, as shown in figure 53.

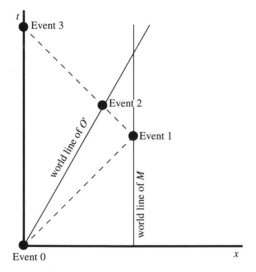

Figure 53

The dashed lines show the world-line of the light beam. We will assume that clocks in both frames are synchronised to $t = t' = 0$ when the origins of the frames coincide at event 0 (emission of the light beam).

The mirror M is a distance d from the origin O, so light will take a time d/c to reach it. Thus

▶ $t_1 = d/c.$

When we perform the Lorentz transformations later we will also need the x-coordinate of this event, which is obviously

▶ $x_1 = d.$

At time t_1, the light beam is at $x = d$. After this time, the light beam travels in the negative x-direction, so at time $t > t_1$ its x-coordinate is

$$x = d - c(t - t_1)$$
$$= d - c(t - d/c)$$
$$= 2d - ct.$$

At time t, the x-coordinate of O' is vt, so at time t_2, when O' and the light beam meet, we must have

$$2d - ct_2 = vt_2.$$

Thus

▶ $$t_2 = \frac{2d}{c + v}.$$

The x-coordinate of this event can be found by substituting into either $x = 2d - ct_2$ or $x = vt_2$, to give

▶ $$x_2 = \frac{2dv}{c + v}.$$

The time coordinate of event 3 is easy to calculate, since it is just the time required for the beam of light to travel from O to M and back again. Thus

▶ $t_3 = 2d/c.$

Clearly

▶ $x_3 = 0.$

Now let us calculate the time-coordinates of these events in the frame S', using the Lorentz transformation

$$t' = \gamma\left(t - \frac{vx}{c^2}\right).$$

Thus

$$t'_1 = \gamma\left(\frac{d}{c} - \frac{vd}{c^2}\right) = \frac{\gamma d}{c}(1 - v/c),$$

$$t'_2 = \gamma\left(\frac{2d}{c+v} - \frac{2v^2 d}{c^2(c+v)}\right) = \frac{2\gamma d}{c+v}(1 - v^2/c^2)$$

(which is equal to $2t'_1$),

$$t'_3 = \gamma\left(\frac{2d}{c} - 0\right) = \frac{2\gamma d}{c}.$$

In order to see how observers in S' would interpret these measurements, it is helpful to draw the space-time diagram for the frame S', as shown in figure 54.

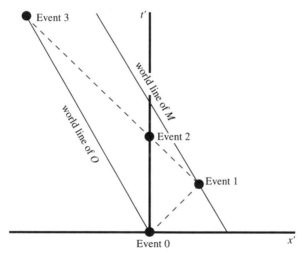

Figure 54

According to the observers in S', O is moving in the negative x'-direction at some speed u. We can use their measurements of t'_1 and t'_3 to calculate u as follows:

At time t', the x'-coordinate of O is

$$x' = -ut'.$$

For times less than t'_1, the light beam moves in the positive x'-direction, and for times greater than t'_1 it moves in the negative x'-direction. Thus at time $2t'_1$ the light beam has an x'-coordinate of zero, and at time $t' > t'_1$ the x'-coordinate is

$$x' = -c(t' - 2t'_1).$$

The x'-coordinates of O and of the light beam must be equal at $t' = t_3'$, so

$$ut_3' = c(t_3' - 2t_1')$$

and hence

$$u = c\left(1 - \frac{2t_1'}{t_3'}\right).$$

Substituting the observed values

$$t_1' = \frac{\gamma d}{c}\left(1 - \frac{v}{c}\right) \quad \text{and} \quad t_3' = \frac{2\gamma d}{c}$$

gives $u = v$, so the observers in S' see the frame S moving backwards at speed v. [We could have written this down straight away, since it follows directly from the postulates of special relativity.]

The observers in S' can also use their measurements to calculate the distance d' between O and M: At time t', the x'-coordinate of M is

$$x' = d' - ut' = d' - vt'.$$

At time t' the x'-coordinate of the light beam is

$$x' = ct',$$

and when $t' = t_1'$ these x'-coordinates must be equal. Thus

$$ct_1' = d' - vt_1',$$

hence

$$d' = (c + v)t_1'.$$

Substituting the observed value of t_1' gives

$$d' = \frac{\gamma d}{c}\left(1 - \frac{v}{c}\right)(c + v) = \gamma d\left(1 - \frac{v^2}{c^2}\right) = d\left(1 - \frac{v^2}{c^2}\right)^{1/2} = \frac{d}{\gamma}.$$

The observers in S' thus measure the distance from O to M to be d/γ, which is in agreement with the length-contraction formula.

Problem 55

A member of a colony on a moon of Jupiter is required to salute the UN flag at the same time as it is being done on Earth, at noon in New York. If observers in all inertial frames are to agree that he has performed his duty, for how long must he salute? (The distance from Earth to Jupiter is 8×10^8 km. The relative motion of the Earth and Jupiter's moon may be ignored.)

Solution

Write down the Lorentz transformation for t in differential form,

$$\Delta t' = \gamma(\Delta t - v\Delta x/c^2),$$

and let S be the frame of reference in which the Earth (and the moon of Jupiter) is at rest, and S' the rest frame of an arbitrary observer.

Identify the salute on Earth and the salute on the moon of Jupiter as two events. In the frame S, the spatial displacement Δx between these two events is D (i.e. the distance from Earth to Jupiter) and the temporal separation Δt is T.

We require that in S' $\Delta t'$ is zero, hence

$$T = vD/c^2.$$

Thus the member of the colony must be saluting at a time vD/c^2 after the salute on Earth, as measured in the frame S, and since v can vary between $\pm c$ he must salute for a total time of

$$
\begin{aligned}
&2D/c \\
&= 2 \times 8 \times 10^{11}/(3 \times 10^8) \text{ s} \\
&= 5.3 \times 10^3 \text{ s} \\
&\approx 1.5 \text{ hours.}
\end{aligned}
$$

▶

Problem 56

Two rockets A and B depart from Earth at steady speeds of $0.6c$ in opposite directions, having synchronised clocks with each other and with Earth at departure. After one year as measured in Earth's reference frame, rocket B emits a light signal. At what times, in the reference frames of the Earth and of rockets A and B, does rocket A receive the signal?

Solution

In this problem we have three inertial frames to consider, so we will use coordinates x and t to denote quantities measured in the Earth's frame, x_A and t_A to denote quantities measured in A's frame, and x_B and t_B to denote quantities measured in B's frame. We will assume that, according to an observer in the Earth's frame, rocket A is travelling at speed v in the positive x-direction and rocket B in the negative x-direction. With these assumptions, we can write down the Lorentz transformations:

$$x = \gamma(x_A + vt_A), \qquad x_A = \gamma(x - vt),$$
$$t = \gamma(t_A + vx_A/c^2), \qquad t_A = \gamma(t - vx/c^2),$$
$$x = \gamma(x_B - vt_B), \qquad x_B = \gamma(x + vt),$$
$$t = \gamma(t_B - vx_B/c^2), \qquad t_B = \gamma(t + vx/c^2),$$

where

$$v = 0.6c$$

and

$$\gamma = (1 - v^2/c^2)^{-1/2} = 5/4.$$

It will be convenient to use a system of units in which time is measured in years and distance in light-years, in which case c has a value of 1. Again, it is helpful to draw a space-time diagram in the Earth's frame of reference, as shown in figure 55.

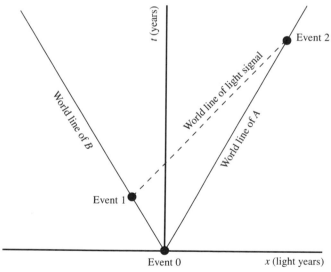

Figure 55

In the Earth's frame of reference, event 1 (B emits the signal) has coordinates

$$x = -0.6,$$
$$t = +1,$$

so at time $t > 1$ the x-coordinate of the light signal is

$$x = -0.6 + (t - 1) = t - 1.6.$$

At time t, the x-coordinate of rocket A is

$$x = +0.6t.$$

The x-coordinates must be equal at event 2, so for this event we must have

$$t - 1.6 = 0.6t,$$

▶ hence $t = 4$. Thus according to an observer in the Earth frame, event 2 occurs after 4 years. The x-coordinate of this event is therefore $0.6 \times 4 = 2.4$.

We can now substitute these coordinates into the Lorentz transformation formulae to find the time coordinates of event 2 in A's and B's frame of reference:

$$t_A = \gamma(t - vx)$$

(remember that $c = 1$)

$$= \frac{5}{4}(4 - 0.6 \times 2.4) = 3.2,$$

▶ so that, according to A, event 2 occurs after 3.2 years.

$$t_B = \gamma(t + vx)$$
$$= \frac{5}{4}(4 + 0.6 \times 2.4) = 6.8,$$

▶ so according to B, event 2 occurs after 6.8 years.

Problem 57

A very fast train of proper length L_0 rushes through a station which has a platform of length L $(< L_0)$. What must be its speed v such that the back of the train is opposite one end of the platform at exactly the same instant as the front of the train is opposite the other end, according to an observer on the platform?

According to this observer, two porters standing at either end of the platform (distance L apart) kick the train simultaneously, thereby making dents in it. When the train stops, the dents are at a distance L_0 apart. How is the difference between L and L_0 explained by (a) the observer on the platform, and (b) an observer travelling in the train?

Solution

According to the observer on the platform, the train undergoes a Lorentz contraction by a factor of γ, so that its length is L_0/γ. This must clearly be equal to L if the two ends of the train are to align with the two ends of the platform, so

$$\gamma = (1 - v^2/c^2)^{-1/2} = L_0/L.$$

Rearranging,

$$v/c = (1 - L^2/L_0^2)^{1/2},$$

so

▶ $$v = c(1 - L^2/L_0^2)^{1/2}.$$

(a) According to the observer on the platform, the kicks are administered at either end of the train so the dents will be at the two ends of the train. When the train stops this is found to be the case, although the train is no longer undergoing a Lorentz contraction so the separation of the dents is greater than it was when the train was in motion.

(b) According to an observer on the train, the train is of length L_0 but the platform is moving at velocity $-v$ so it undergoes Lorentz contraction from its proper length L to a length $L/\gamma = L^2/L_0$. The fact that two porters standing this distance apart nevertheless manage to make dents in the train separated by L_0 is explained by the fact that the kicks are not administered simultaneously.

We can show this using the Lorentz transformations. Let us identify frame S as the frame in which the platform is stationary, and S' as the frame in which the train is stationary. The train thus has a velocity $+v$ in the x-direction relative to the platform. We will call the two porters A and B, and assume that their x-coordinates are 0 and L respectively in the frame S. We will also assume that, in frame S, the kicks occur at time $t = 0$. Thus we have, using the Lorentz transformations, the data shown in Table 5.

Table 5

	In frame S	In frame S'
A kicks the train	$x = 0$, $t = 0$	$x' = 0$, $t' = 0$
B kicks the train	$x = L$, $t = 0$	$x' = \gamma L$, $t' = -\gamma v L/c^2$

In order for an observer at rest in the frame S' to measure the distance between the kicks as L_0, we must have $\gamma = L_0/L$ as before. However, we

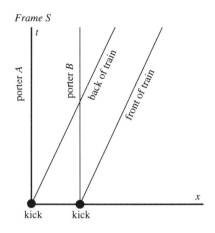

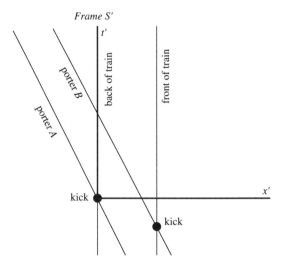

Figure 56

see that in this frame porter B's kick is administered earlier than porter A's kick, by an interval of time equal to $\gamma v L/c^2$. To show that this is consistent with the Lorentz contraction of the platform observed by the train, we can use this time interval to calculate the length of the platform according to an observer in S'.

According to observers in S', the porters are moving at speed v in the negative x'-direction, so at $t' = 0$, porter B is at $x' = L_0 - \gamma v^2 L/c^2$ and porter A is at $x' = 0$. The length of the platform, according to an observer in S', is thus $L_0 - \gamma v^2 L/c^2$. Using the fact that $\gamma = L_0/L$, this can be rewritten as $L_0(1 - v^2/c^2)$, and using the fact that $(1 - v^2/c^2) = 1/\gamma^2$, it can be simplified to L_0/γ^2. Thus according to an observer moving with the train, the length of the platform is $L_0/\gamma^2 = L/\gamma$, so it has been Lorentz-contracted as we expect.

This can be summarised on space-time diagrams in S and S', shown in figure 56.

Problem 58

Given two observers O and O', with O' moving at uniform velocity v in the positive x-direction relative to O, use the appropriate Lorentz transformations to show that if an object is moving with velocity component $u_{x'}$ in the frame of reference of O', then

$$u_x = \frac{u_{x'} + v}{1 + \dfrac{v u_{x'}}{c^2}},$$

where u_x is the corresponding velocity component according to O.

(a) A space ship is launched from Earth and maintains a uniform velocity of $0.900c$. The space ship subsequently launches a small rocket in the forward direction with a speed of $0.900c$ relative to the ship. What is the speed of the small rocket relative to the Earth?

(b) According to observations on the Earth, the nearest star to the solar system is 4.25 light years away. A space ship which leaves the Earth and travels at uniform velocity takes 4.25 years, according to ship-borne clocks, to reach the star. What is the speed of the space ship, expressed as a fraction of the speed of light c?

Solution

The frames of reference of O and O' have the standard relationship, so that the Lorentz transformations are given, in differential form, by

$$dx' = \gamma(dx - v\,dt),$$
$$dt' = \gamma\left(dt - \frac{v\,dx}{c^2}\right).$$

[We could also write down the reverse transformations, but in fact we don't need them.] The ratio of these two expressions gives the component $u_{x'}$:

$$u_{x'} = \frac{dx'}{dt'} = \frac{dx - v\,dt}{dt - \dfrac{v\,dx}{c^2}}.$$

Multiplying through by the denominator of the right-hand side and rearranging, we obtain

$$dx\left(1 + \frac{vu_{x'}}{c^2}\right) = dt(u_{x'} + v),$$

from which we can write down the required result:

▶ $$u_x = \frac{dx}{dt} = \frac{u_{x'} + v}{1 + \dfrac{vu_{x'}}{c^2}}.$$

(a) This result can be applied directly if we identify O with an observer on Earth and O' with an observer on the space ship. We have $v = 0.900c$ and $u_{x'} = 0.900c$, so

$$u_x = \frac{0.900c + 0.900c}{1 + \dfrac{(0.900c)(0.900c)}{c^2}} = \frac{1.800c}{1.810}.$$

▶   The speed of the rocket relative to Earth is thus $0.994c$.

(b) We can solve this using the Lorentz transformations. Let us identify S as the frame in which the Earth and the star are at rest, and S' as the frame in which the space ship is at rest, and synchronise clocks to $t = t' = 0$ when $x = x' = 0$. If we put D for the distance to the star in the frame S, the coordinates in S of the event of the space ship reaching the star are

$$x = D,$$
$$t = D/v.$$

Thus the time coordinate of this event in the space ship's frame S' is

$$\begin{aligned} t' &= \gamma(t - vx/c^2) \\ &= \gamma(D/v - Dv/c^2) \\ &= (\gamma D/v)(1 - v^2/c^2) \\ &= D/\gamma v. \end{aligned}$$

Now we are given that $t' = 4.25$ years and $D/c = 4.25$ years, so it follows that

$$\gamma v/c = 1.$$

Putting

$$\beta = v/c$$

for convenience, we have

$$\gamma\beta = 1,$$

and since

$$\gamma = \frac{1}{\sqrt{(1 - \beta^2)}},$$

this gives

$$\frac{\beta^2}{1 - \beta^2} = 1,$$

so that

$$\beta^2 = 1/2,$$

▶ hence $v = c/\sqrt{2}$.

Problem 59

To an observer, two bodies of equal rest mass collide head on with equal but opposite velocities $4c/5$ and cohere. To a second observer, one body is initially at rest. Find the apparent velocity of the other, moving mass before the collision and compare its initial energy in the two frames of reference.

Solution

It is clear that the apparent velocity of the second mass must be equal to the relative velocity of the two masses, which is given by the relativistic addition of $4c/5$ and $4c/5$:

$$v = \frac{4c/5 + 4c/5}{1 + (4c/5)(4c/5)(1/c^2)} = \frac{8c/5}{41/25}$$

▶ $$= 40c/41.$$

The total initial energy of either particle in the first frame is γmc^2, where m is its rest mass and γ is the Lorentz factor for a speed of $4c/5$, thus

$$E = \frac{mc^2}{\sqrt{\left(1 - \frac{4^2}{5^2}\right)}}$$

$$= (5/3)mc^2.$$

In the second frame, the same formula applies for the moving particle but γ is now the value appropriate to a speed of $40c/41$, thus

$$E = \frac{mc^2}{\sqrt{\left(1 - \dfrac{40^2}{41^2}\right)}}$$
$$= (41/9)mc^2.$$

Thus the initial energy of the moving particle is greater by a factor of ▶ $41/15$ in the second frame.

Problem 60

A beam of monochromatic light, whose wavelength in free space is λ, is split into two separate beams and each is then passed through identical troughs of water. Show that if the water in one trough is stationary and the water in the other trough is moving with speed v ($\ll c$) in the direction of the light, the phase difference between the emerging beams is

$$\Delta\phi = (2\pi L/\lambda)(n^2 - 1)(v/c),$$

where L is the length of the troughs and n is the refractive index of the stationary water. Suggest suitable values for L and v in an experimental arrangement for verifying this result.

Solution

Figure 57 shows the arrangement.

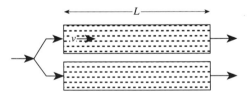

Figure 57

The refractive index n defines the speed at which light propagates with respect to the medium as c/n, and since the moving water moves forward (i.e. in the same direction as the light) in the laboratory frame at speed v, the speed of the light relative to the laboratory frame must be the relativistic sum of v and c/n, i.e.

$$\frac{v + \dfrac{c}{n}}{1 + \dfrac{v}{nc}}.$$

Thus the time taken for light to travel from one end of the trough of moving water to the other, measured in the laboratory frame, is

$$T_{\text{moving}} = L(1 + v/nc)/(c/n + v)$$
$$= (Ln/c)(1 + v/nc)(1 + vn/c)^{-1}.$$

Since $v \ll c$, we can use the binomial theorem to expand this expression to the first order in v, giving

$$T_{\text{moving}} \approx (Ln/c)(1 + v/nc - vn/c).$$

The time required for light to travel through the trough of stationary water can be deduced from this expression by substituting into it $v = 0$, to give

$$T_{\text{stationary}} = Ln/c$$

(which is obviously correct), so the difference in travel times through the two troughs is

$$\Delta T = (Ln/c)(v/c)(n - 1/n)$$
$$= (Lv/c^2)(n^2 - 1).$$

The phase difference $\Delta\phi$ is given by $2\pi v \Delta T$ where v is the frequency, and substituting $v = c/\lambda$ finally gives

▶ $$\Delta\phi = (2\pi L/\lambda)(n^2 - 1)(v/c)$$

as required.

Water has a refractive index n of approximately 1.33 at optical wavelengths (say 500 nm), and if we assume that $\Delta\phi$ must be at least $\pi/2$ to produce a measurable effect, substitution into this expression shows that Lv must exceed about 50 m^2 s^{-1}. Possible values for a demonstration might be $L = 5$ m and $v = 10$ m s^{-1}.

Problem 61

In its rest frame, a source emits light in a conical beam of width $\pm 45°$. In a frame moving towards the source at speed v, the beam width is $\pm 30°$. What is v?

Solution

There are several ways of solving this. The longest, but most basic, is to use the Lorentz transformations directly. It is also possible to use the formulae for relative velocity, or (the shortest method) to use the aberration formula.

(1) Using the Lorentz transformation directly.

First we need to set up two inertial frames S and S' in which to describe the problem. If we adopt the standard configuration in which frame S' has a positive velocity v in the x-direction when observed in frame S, figure 58 shows that we can consider the light to be emitted at up to $\pm 45°$ from the x'-axis in the frame S'.

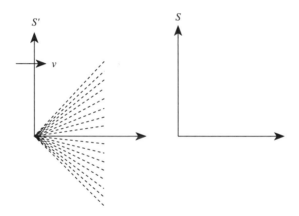

Figure 58

We can write the spatial parts of the Lorentz transformations in differential form:

$$\Delta x = \gamma(\Delta x' + v\Delta t'),$$
$$\Delta y = \Delta y'.$$

If we consider a photon travelling at the very edge of the cone, in S' it will make an angle of 45° with the x'-axis. If the photon connects two events which have a separation along the x'-axis of $\Delta x'$, the separation along the y'-axis must be numerically equal to $\Delta x'$ (because the photon is travelling at an angle of 45°), and the time interval $\Delta t'$ must be $\sqrt{2}\Delta x'/c$ (because the photon travels a distance $\sqrt{2}\,\Delta x'$ at the speed of light).

Thus we have

$$\Delta x',$$
$$\Delta y' = \Delta x',$$
$$\Delta t' = \sqrt{2}\,\Delta x'/c.$$

Transforming into the frame S, we find

$$\Delta x = \gamma\left(\Delta x' + \frac{\sqrt{2}\,v\Delta x'}{c}\right) = \gamma\Delta x'\left(1 + \frac{\sqrt{2}\,v}{c}\right),$$
$$\Delta y = \Delta x'.$$

The photon must make an angle of $\arctan(\Delta y/\Delta x)$ with the x-axis (which we are told is 30°), so

$$\gamma\left(1 + \frac{\sqrt{2}\,v}{c}\right) = \frac{1}{\tan 30°} = \sqrt{3}.$$

Substituting $\gamma = (1 - v^2/c^2)^{-1/2}$ gives

$$\frac{1 + \sqrt{2}(v/c)}{\sqrt{(1 - (v/c)^2)}} = \sqrt{3}.$$

If we put $\beta = v/c$ for convenience, and multiply throughout by $(1 - \beta^2)^{1/2}$, we obtain

$$1 + \sqrt{2}\,\beta = (3 - 3\beta^2)^{1/2}.$$

Squaring:

$$1 + 2\sqrt{2}\,\beta + 2\beta^2 = 3 - 3\beta^2.$$

This can be rearranged as a quadratic equation in β:

$$5\beta^2 + 2\sqrt{2}\,\beta - 2 = 0.$$

Solving the quadratic gives $v/c = +0.410$ or -0.976. Clearly we require the positive solution, so our result is $v = 0.41\,c$. [The negative solution corresponds to the transformed cone making an angle of $-30°$ with the axis. It was introduced when we squared our expression for β.]

(2) Using the velocity transformation formulae.

A photon travelling along the edge of the cone has velocity components, in S', of

$$v_x' = c/\sqrt{2}$$

and

$$v_y' - c/\sqrt{2}.$$

In the frame S, these will transform to v_x and v_y where, in general,

$$v_x = \frac{v + v'_x}{1 + vv'_x/c^2}$$

and

$$v_y = \frac{v'_y(1 - v^2/c^2)^{1/2}}{1 + vv'_x/c^2}.$$

In this particular case, we thus obtain

$$v_x = \frac{v + \dfrac{c}{\sqrt{2}}}{1 + \dfrac{v}{\sqrt{2}\,c}},$$

$$v_y = \frac{c}{\sqrt{2}} \frac{\sqrt{\left(1 - \dfrac{v^2}{c^2}\right)}}{1 + \dfrac{v}{\sqrt{2}\,c}}.$$

Since $v_y/v_x = \tan 30° = 1/\sqrt{3}$ we have

$$\frac{1 + \sqrt{2}(v/c)}{\sqrt{(1 - (v/c)^2)}} = \sqrt{3}$$

as before.

[If we were unable to remember the transformation for v_y, we could solve the problem using the velocity transformation for v_x alone, using the fact that the speed of light is the same in all inertial frames, as shown in figure 59.

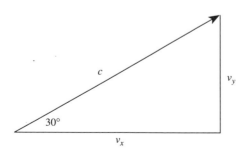

Figure 59

Thus $v_x/c = \cos 30° = \sqrt{3}/2$, so

$$\frac{v + \dfrac{c}{\sqrt{2}}}{1 + \dfrac{v}{\sqrt{2}\,c}} = \frac{\sqrt{3}\,c}{2}.$$

Cross-multiplying this expression gives

$$2v + \sqrt{2}\,c = \sqrt{3}\,c + \sqrt{(3/2)}\,v,$$

which can be solved to give

$$v = c\frac{\sqrt{3} - \sqrt{2}}{2 - \sqrt{\dfrac{3}{2}}}$$

$= 0.410\,c$ as before.]

(3) Using the aberration formula for light making an angle ϕ with the x-axis and ϕ' with the x'-axis.

$$\tan{(\phi/2)} = \sqrt{\frac{1 - v/c}{1 + v/c}}\,\tan{(\phi'/2)}.$$

Substitution of $\phi = 30°$ and $\phi' = 45°$ gives

$$\frac{1 - v/c}{1 + v/c} = \frac{\tan^2 15°}{\tan^2 22.5°} = 0.4185,$$

therefore

$$\frac{v}{c} = \frac{1 - 0.4185}{1 + 0.4185}$$
$$= 0.41.$$

Problem 62

Estimate the minimum frequency of a γ-ray that causes a deuteron to disintegrate into a proton and a neutron, commenting on any assumptions you make. The masses of the particles are

$$m_d = 2.0141 m_u,$$
$$m_p = 1.0078 m_u,$$
$$m_n = 1.0087 m_u.$$

Solution

We will assume that the deuteron is at rest, and that the proton and neutron are created at rest. This cannot be quite correct, since it violates the principle of conservation of momentum, but the masses of the particles involved are large so the associated velocities will be small.

The total mass of the products is $2.0165m_u$, which is greater than the mass of the deuteron by $0.0024m_u$. The extra mass must be provided by the energy of the photon, so the minimum possible frequency must be given by

$$hv = 0.0024m_u c^2.$$

▶ Substituting $m_u = 1.66 \times 10^{-27}$ kg gives $v = 5.4 \times 10^{20}$ Hz.

[We can roughly check the reliability of our assumption as follows. The momentum of this photon is given by $hv/c = p = 1.2 \times 10^{-21}$ kg m s^{-1}. If all of this momentum were transferred to (say) the proton, it would acquire a velocity of $p/m = 7.2 \times 10^{5}$ m s^{-1}. This is much less than the speed of light, so we are justified in ignoring changes in mass caused by the velocities of the particles. In fact, the error in our calculation can be shown to be less than 0.1%.]

Problem 63

What is the speed of an electron which has a total energy of 1 MeV?

Solution

The total energy E is given by

$$E = \gamma m_0 c^2,$$

so

$$
\begin{aligned}
\gamma &= E/(m_0 c^2) \\
&= 10^6 \times 1.60 \times 10^{-19}/(9.11 \times 10^{-31})/(3.00 \times 10^8)^2 \\
&= 1.95.
\end{aligned}
$$

Now

$$\gamma = (1 - v^2/c^2)^{-1/2},$$

so

$$
\begin{aligned}
v/c &= (1 - 1/\gamma^2)^{1/2} \\
&= 0.86.
\end{aligned}
$$

Therefore

▶ $v = 0.86c.$

Problem 64

A particle of rest mass m_0 is travelling so that its total energy is just twice its rest mass energy. It collides with a stationary particle of rest mass m_0 to form a new particle. What is the rest mass of the new particle?

Solution

Before the collision, particle 1 (the moving particle) has a total energy of $2m_0c^2$ and a non-zero momentum which we will call p, as shown in figure 60. Particle 2 has a total energy of m_0c^2 and a momentum of zero. Thus by conservation of energy and momentum, the new particle has a total energy of $3m_0c^2$ and a total momentum of p.

Before

particle 1   particle 2

p

m_0 m_0

After

new particle

p

m_1

Figure 60

We can find p by using the energy–momentum invariant for particle 1:

$$E^2 - p^2c^2 = m_0^2c^4;$$

rearranging,

$$\begin{aligned}
p^2 &= E^2/c^2 - m_0^2c^2 \\
&= (2m_0c^2)^2/c^2 - m_0^2c^2 \\
&= 3m_0^2c^2.
\end{aligned}$$

Applying the same invariant to the new particle (whose rest mass we will call m_1), we have

$$E^2 - p^2c^2 = m_1^2c^4,$$

so

$$\begin{aligned}
m_1^2c^4 &= (3m_0c^2)^2 - 3m_0^2c^2 \cdot c^2 \\
&= 6m_0^2c^4.
\end{aligned}$$

Thus

▶ $m_1 = \sqrt{6}\, m_0.$

Problem 65

Explain carefully why an uncharged pi-meson (mass 134 MeV/c^2) can always decay into two photons whereas a photon of sufficient energy can decay into an electron–positron pair only in the presence of matter.

Solution

(1) Decaying pi-meson.

However the pi-meson is moving, we can define its zero momentum frame (ZMF) in which it is at rest. In this frame it can clearly decay into two photons of equal and opposite momentum, each of which carries a total energy of 67 MeV. If the process is possible in one inertial frame, it must be possible in all inertial frames even though the details (energy and momentum of the photons) will differ.

(2) Decaying photon.

If we assume this to be possible in the absence of matter, the electron–positron pair produced by the decay will have a ZMF. However, a photon cannot have zero momentum, so the decay process is impossible in the ZMF and hence in all frames.

   In the presence of another particle, however, we can allocate energy and momentum between the photon and the particle as required, and the process becomes possible. For example, let us consider a photon of energy $h\nu$ decaying in the presence of a particle of rest mass M to produce an electron–positron pair (each of rest mass m), and the original particle of mass M, all at rest (see figure 61).

   Since all the particles after the decay are at rest, the total momentum of the system is zero. Since the photon's momentum is $+h\nu/c$, the initial momentum of the particle of mass M must be $-h\nu/c$.

   The total energy of the system after the decay is $(M + 2m)c^2$, which must be equal to the total energy before the decay. Since the energy of the photon is $h\nu$, the initial energy of the particle of mass M must be $(M + 2m)c^2 - h\nu$.

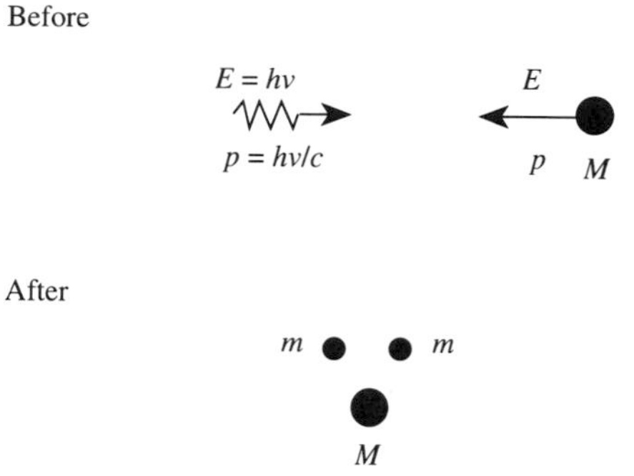

Before

$E = hv$

$p = hv/c$

E

p M

After

m ● ● m

●

M

Figure 61

We know that for a single particle of rest mass M, the energy E and momentum p are related by

$$E^2 - p^2c^2 = M^2c^4,$$

so on substituting for E and p we obtain

$$(M + 2m)^2c^4 + h^2v^2 - 2(M + 2m)c^2hv - h^2v^2 = M^2c^4.$$

Therefore

$$hv = \frac{(M + 2m)^2c^4 - M^2c^4}{2(M + 2m)c^2}$$

$$= \frac{2mc^2(1 + m/M)}{(1 + 2m/M)}.$$

Thus there is a solution for any value of M, and we deduce that the decay is possible if the photon has sufficient energy.

Problem 66

A particle of rest mass m moving along the x-axis with velocity v collides with a particle of rest mass $m/2$ moving along the x-axis with velocity $-v$. If the two particles coalesce, find the rest mass of the resulting particle.

Solution

Figure 62 shows the situation before and after the collision. Particle 1 (the one with mass m) has a total energy of γmc^2, where $\gamma = (1 - v^2/c^2)^{-1/2}$, and particle 2 has a total energy of $0.5\gamma mc^2$, so by conservation of energy the resulting particle must have a total energy of $1.5\gamma mc^2$.

Figure 62

Particle 1 has a momentum of $\gamma m v$, and particle 2 has a momentum of $-0.5\gamma m v$, so by conservation of momentum the resulting particle must have a momentum of $0.5\gamma m v$.

We now know the energy and momentum of the resulting particle, so we can use the energy–momentum invariant to find its rest mass m_1:

$$E^2 - p^2 c^2 = m_1^2 c^4,$$

so

$$\begin{aligned} m_1^2 &= E^2/c^4 - p^2/c^2 \\ &= (9/4)\gamma^2 m^2 - (1/4)\gamma^2 m^2 v^2/c^2 \\ &= (1/4)\gamma^2 m^2 (9 - v^2/c^2). \end{aligned}$$

Substituting the expression for $\gamma = (1 - v^2/c^2)^{-1/2}$, we finally obtain

$$m_1 = \frac{m}{2}\sqrt{\frac{9 - v^2/c^2}{1 - v^2/c^2}}.$$

As a simple check, we can see that when $(v/c)^2 \ll 1$ this tends to $3m/2$, which is clearly the correct classical limit.

Problem 67

The proton collider at CERN in Geneva makes use of proton and antiproton beams travelling in opposite directions. Explain the advantages of this technique over that of using an antiproton beam hitting stationary protons by calculating the minimum energy of the antiprotons ($\bar{p}$) needed to give the following reaction in which $\Omega\bar{\Omega}$ particle–antiparticle pairs are produced:

$$p + \bar{p} \rightarrow p + \bar{p} + \Omega + \bar{\Omega}:$$

(a) for colliding antiproton and proton beams;
(b) for antiprotons hitting stationary protons.

Express your answers in terms of the proton rest-mass energy.

(The Ω has a mass of $1.78m_p$.)

Solution

(a) Figure 63 shows the situation before and after the collision. If the proton and antiproton collide with equal and opposite velocities, the laboratory frame is the zero momentum frame (ZMF). The resulting system of particles thus has no net momentum, so the configuration of minimum energy is when all four particles are at rest. The total energy after the collision is thus

$$2m_p c^2 + 2m_\Omega c^2 = 5.56 m_p c^2.$$

By conservation of energy, this must be equal to the total energy of the two particles before the collision, so by symmetry the energy of the antiproton must be half of this value, i.e. $2.78 m_p c^2$.

Before

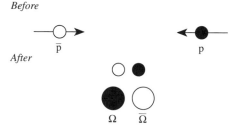

$\bar{p}$ p

After

Ω $\bar{\Omega}$

Figure 63

(b) If the proton is initially at rest, the description in part (a) is still valid except that we need to transform it into a different inertial frame, as shown in figure 64.

In ZMF

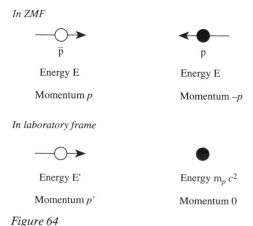

$\bar{p}$ p

Energy E   Energy E

Momentum p Momentum $-p$

In laboratory frame

Energy E'   Energy $m_p c^2$

Momentum p' Momentum 0

Figure 64

In the zero momentum frame, the system has a total energy of $2E$ and a total momentum of zero. In the laboratory frame, it has a total energy of $E' + m_p c^2$ and a total momentum of p'. Now for any system, the quantity

$$E^2 - p^2 c^2$$

is a Lorentz invariant, i.e. it is the same in any inertial frame, so we have the relationship

$$(2E)^2 = (E' + m_p c^2)^2 - p'^2 c^2,$$

which can be expanded as

$$4E^2 = E'^2 + m_p^2 c^4 + 2E' m_p c^2 - p'^2 c^2.$$

However, the energy E' and the momentum p' of the antiproton are related by

$$E'^2 - p'^2 c^2 = m_p^2 c^4,$$

and we can substitute this result to eliminate p':

$$2E^2 = m_p^2 c^4 + E' m_p c^2.$$

Therefore

$$E' = (2E^2 - m_p^2 c^4)/m_p c^2.$$

Now E is the value that we calculated in part (a), and we want to express E' in terms of the proton rest-mass energy, so if we divide this expression by $m_p c^2$ we obtain

$$
\begin{aligned}
\frac{E'}{m_p c^2} &= 2 \left(\frac{E}{m_p c^2} \right)^2 - 1 \\
&= 2(2.78)^2 - 1 \\
&= 14.5.
\end{aligned}
$$

Thus the minimum antiproton energy if the proton is stationary is $14.5 m_p c^2$.

Problem 68

A proton of total energy E collides elastically with a second proton at rest in the laboratory. After the collision the two protons follow trajectories which are disposed symmetrically at angles $\pm \phi/2$ to the direction of the incident particle. By considering the motion in the laboratory frame, or otherwise, show that

$$\cos\phi = \frac{E - E_0}{E + 3E_0},$$

where E_0 is the rest mass energy of the proton.

What is the value of ϕ when the first proton is accelerated from rest through a potential difference of 1.5×10^9 V before colliding with the second proton?

Solution

Let us call the momentum of the initially moving particle p_1, and the energy and (modulus) momentum of each particle after the collision E_2 and p_2 respectively, as shown in figure 65.

Before

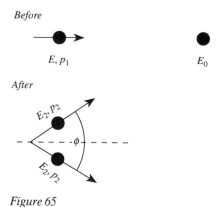

After

Figure 65

Conservation of momentum gives

$$p_1 = 2p_2 \cos(\phi/2).$$

The relationship between E and p_1 is

$$E^2 - p_1^2 c^2 = E_0^2,$$

which we can substitute into our expression to obtain

$$(E^2 - E_0^2) = 4p_2^2 c^2 \cos^2(\phi/2).$$

We can eliminate p_2 from this by using the energy–momentum relation again:

$$E_2^2 - p_2^2 c^2 = E_0^2,$$

which gives

$$(E^2 - E_0^2) = 4(E_2^2 - E_0^2)\cos^2(\phi/2).$$

Finally we can eliminate E_2 using the principle of conservation of energy:

$$E + E_0 = 2E_2,$$

which gives

$$(E^2 - E_0^2) = (E^2 - 3E_0^2 + 2EE_0)\cos^2(\phi/2),$$

so

$$\cos^2\left(\frac{\phi}{2}\right) = \frac{E^2 - E_0^2}{E^2 + 2EE_0 - 3E_0^2}.$$

Recalling that $\cos\phi = 2\cos^2(\phi/2) - 1$, we can rearrange this to give

$$\begin{aligned}\cos\phi &= \frac{E^2 + E_0^2 - 2EE_0}{E^2 + 2EE_0 - 3E_0^2}\\ &= \frac{(E - E_0)^2}{(E - E_0)(E + 3E_0)}.\end{aligned}$$

Hence

$$\blacktriangleright \qquad \cos\phi = \frac{E - E_0}{E + 3E_0}$$

as required.

If the first proton is accelerated through a potential V, it acquires a kinetic energy of eV, so

$$E = E_0 + eV;$$

therefore

$$\begin{aligned}E/E_0 &= 1 + eV/E_0\\ &= 1 + eV/mc^2\\ &= 1 + 1.60 \times 10^{-19} \times 1.5 \times 10^9/(1.67 \times 10^{-27})/(3.00 \times 10^8)^2\\ &= 2.60,\end{aligned}$$

so

$$\begin{aligned}\cos\phi &= (2.60 - 1)/(2.60 + 3)\\ &= 0.286;\end{aligned}$$

$\blacktriangleright$ therefore $\phi = 73°$.

Problem 69

Consider the elastic scattering of a photon of frequency v by a stationary electron (the Compton effect). Find an expression for the wavelength change of a photon scattered through 180°. What is the energy of a photon of initial energy 1 MeV after a single 180° scattering?

Solution

Let us assume that the photon has a frequency v' after being scattered, and that the electron acquires a momentum p as a result of the collision, as shown in figure 66.

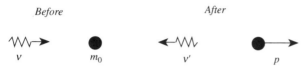

Before   After

v m_0 v' p

Figure 66

Since the momentum of a photon of frequency v is given by hv/c, conservation of momentum gives

$$\frac{hv}{c} = p - \frac{hv'}{c}, \tag{1}$$

and since the energy of a particle of rest mass m_0 and momentum p is given by $(m_0^2 c^4 + p^2 c^2)^{1/2}$, conservation of energy gives

$$hv + m_0 c^2 = hv' + (m_0^2 c^4 + p^2 c^2)^{1/2}. \tag{2}$$

Rearranging (2) to separate the square root gives

$$(m_0^2 c^4 + p^2 c^2)^{1/2} = hv - hv' + m_0 c^2,$$

and squaring this gives

$$m_0^2 c^4 + p^2 c^2 = h^2 v^2 + h^2 v'^2 + m_0^2 c^4 - 2h^2 vv' + 2hv m_0 c^2$$
$$- 2hv' m_0 c^2. \tag{3}$$

We can eliminate p from this by rearranging (1),

$$pc = hv + hv',$$

and squaring,

$$p^2 c^2 = h^2 v^2 + h^2 v'^2 + 2h^2 vv'.$$

time on the mirror must be

$$\frac{P}{hv}.$$

Now the momentum of each photon is $h/\lambda - hv/c$, so the change in momentum of a single photon when it is reflected from the mirror must be $2hv/c$. Thus the rate at which the momentum of the beam of photons is being changed, which must (by Newton's third law) be equal to the force on the mirror, is

$$F = \frac{P}{hv}\frac{2hv}{c} = \frac{2P}{c}.$$

[We see that neither the frequency of the photons nor Planck's constant appears in the answer, so we do not need to know them. This suggests, correctly, that the result could have been derived from the wave theory of light.] Substituting $P = 10\,\text{mW}$ gives $F = 6.7 \times 10^{-11}\,\text{N}$.

Problem 72

A crystalline specimen is irradiated with X-rays of wavelength λ. Bragg reflexion is observed at 23.0°, ϕ and 73.5° (23.0° $< \phi <$ 73.5°), with respect to the direction of the incident beam. Calculate a possible value for ϕ.

Solution

Bragg reflexion occurs when beams reflected from adjacent crystal planes, whose separation is d, are reflected in phase.

The path difference between the two beams shown in figure 67 is $2d \cos \theta$, so for the beams to be reflected in phase,

$$2d \cos \theta = n\lambda,$$

where n is an integer.

Since the angles specified in the problem are the angles of the reflected beams with respect to the incident beams, they correspond to the angles $180° - 2\theta$. Thus the two known values of θ are 78.5° and 53.25°. In order to find the values of n to which these are likely to correspond, we take the ratio of the cosines of the angles:

$$\frac{\cos 53.25}{\cos 78.5} = 3.00.$$

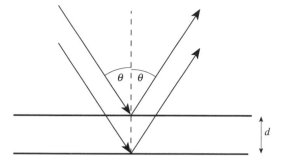

Figure 67

Thus the simplest solution to the problem is that the beam with $\theta = 78.5°$ corresponds to $n = 1$, and that with $\theta = 53.25°$ to $n = 3$ (although $n = 2$ and 6 respectively, or 3 and 9, and so on, would also be possible). The beam with $2\theta = 180° - \phi$ must therefore correspond to $n = 2$, so

$$\frac{\cos(90° - \phi/2)}{2} = \frac{\cos 53.25°}{3}.$$

▶ This gives $\phi = 47.0°$.

Problem 73

In an experiment on the photoelectric effect it is observed that for light of wavelength 500 nm, a stopping potential of 0.25 V is required to cut off the current of photoelectrons, whereas at a wavelength of 375 nm a stopping potential of 1.0 V is required. Calculate from these data the ratio of Planck's constant to the electronic charge (h/e).

Solution

The relationship between the stopping potential V_S and the wavelength λ is

$$eV_S = \frac{hc}{\lambda} - \Phi,$$

where c is the speed of light and Φ is the work function energy of the material being illuminated. Dividing this equation by e and calculating $V_{S1} - V_{S2}$, the difference in stopping potential for two wavelengths λ_1 and λ_2, gives

$$V_{S1} - V_{S2} = \frac{hc}{e}\left(\frac{1}{\lambda_1} - \frac{1}{\lambda_2}\right).$$

Thus

$$\frac{h}{e} = \frac{V_{S1} - V_{S2}}{c(1/\lambda_1 - 1/\lambda_2)}.$$

Taking $V_{S1} = 1.0$ V, $V_{S2} = 0.25$ V, $\lambda_1 = 375$ nm, $\lambda_2 = 500$ nm and
▶ $c = 2.998 \times 10^8$ m s^{-1} gives $h/e = 3.75 \times 10^{-15}$ J s C^{-1}.

The result is not particularly close to the currently accepted value of 4.136×10^{-15} J s C^{-1}, so we ought to estimate its uncertainty. If we assume that the data have been specified to no more significant figures than are justified by the experimental errors, these errors must correspond to at least ± 1 in the least significant figures. Thus $V_{S1} = 1.0 \pm 0.1$ V etc. It is clear that the errors in λ are insigificant compared with the errors in V_S, and that the error in $V_{S1} - V_{S2}$ is dominated by the error of ± 0.1 V in V_{S1}. Thus $V_{S1} - V_{S2} \approx 0.75 \pm 0.1$ V, which is a fractional error of 13%. We would therefore be justified in
▶ quoting the experimental value of h/e as $(3.8 \pm 0.5) \times 10^{-15}$ J s C^{-1}. This range includes the currently accepted value.

Problem 74

What is the velocity of an electron which has a kinetic energy equal to the energy of a photon of wavelength 1 nm?

Solution

A photon of wavelength λ has a frequency c/λ and thus an energy hc/λ. The kinetic energy of a particle of rest mass m is $(\gamma - 1)mc^2$, where γ is defined by

$$\gamma = \frac{1}{\sqrt{(1 - v^2/c^2)}}.$$

Thus

$$\gamma = 1 + \frac{h}{\lambda mc}$$
$$= 1.00243.$$

Now from the definition of γ, we can derive

$$v = c\sqrt{\left(1 - \frac{1}{\gamma^2}\right)};$$

▶ hence $v = 2.1 \times 10^7$ m s^{-1}.

[In fact, the electron velocity is non-relativistic so we could have obtained our answer rather more easily by putting

$$\frac{hc}{\lambda} = \frac{1}{2}mv^2.]$$

Problem 75

To study crystal diffraction we need wavelengths of about 0.5×10^{-10} m. What would be the corresponding kinetic energies in eV of (a) a photon, (b) an electron, and (c) a neutron?

Solution

(a) For a photon, $E = h\nu = hc/\lambda$.
Substituting the value for λ gives $E = 4.0 \times 10^{-15}$ J.
Divide by the electron charge 1.6×10^{-19} C to convert this to electron volts: 25 keV.

(b) and (c) For the particles, assuming the velocities are non-relativistic, we can write the kinetic energy as

$$E = \frac{1}{2}mv^2 = \frac{p^2}{2m},$$

where p is the momentum and m is the particle mass. Using the de Broglie relation

$$p = \frac{h}{\lambda}$$

to relate momentum and wavelength gives

$$E = \frac{h^2}{2m\lambda^2}.$$

(b) For the electron, $m = 9.1 \times 10^{-31}$ kg so $E = 9.6 \times 10^{-17}$ J = 0.60 keV.

(c) For the neutron, $m = 1.68 \times 10^{-27}$ kg so $E = 5.2 \times 10^{-20}$ J = 0.33 eV.

Both of these values are very much lower than the corresponding rest-mass energies, so our assumption that the particles are non-relativistic is justified.

Problem 76

Estimate the speed and the de Broglie wavelength of an oxygen molecule at room temperature.

Solution

The mean square speed $\langle v^2 \rangle$ is given by

$$\frac{1}{2}m\langle v^2 \rangle = \frac{3}{2}kT,$$

where m is the mass of a molecule, k is Boltzmann's constant and T is the absolute temperature. Since the relative molecular mass of the O_2 molecule is 32.0, $m = 32.0 \times 1.66 \times 10^{-27}$ kg $= 5.31 \times 10^{-26}$ kg. We may take room temperature to be 20 °C = 293 K, so $\langle v^2 \rangle = 2.28 \times 10^5$ m^2 s^{-2}, ▶ hence the root mean square speed is 4.8×10^2 m s^{-1}.

The de Broglie wavelength $\lambda = h/p$ where h is Planck's constant and p is the momentum, and since the speed is clearly non-relativistic we may put

$$p = mv.$$

▶ Thus $\lambda = 2.6 \times 10^{-11}$ m = 26 pm.

Problem 77

Use the de Broglie relation between the wavelength and the linear momentum of a non-relativistic particle, and the Planck relation between frequency and quantum energy, to obtain the phase and group velocities of the wave motion associated with a particle with velocity $v \ll c$.

A de Broglie wave travelling in the z-direction passes through a narrow slit, of width a in the x-direction and long in the y-direction. Derive a relationship between the width of the slit and the angular spread of the emerging beam in the xz-plane, and show that it is consistent with the uncertainty principle.

Solution

The de Broglie relation between the linear momentum p of a particle and its wavelength λ is

$$\lambda = \frac{h}{p},$$

although it will be more useful to write it as

$$p = \frac{hk}{2\pi},$$

where k is the wavenumber of the de Broglie wave. The Planck relation between the energy E and the angular frequency ω is

$$E = \frac{h\omega}{2\pi}.$$

To find the phase and group velocities of the wave we need to know its dispersion relation, i.e. the relationship between ω and k, and this is clearly equivalent to knowing the relationship between E and p. Since for a non-relativistic particle the kinetic energy E of a particle of mass m and velocity v is $mv^2/2$ and the momentum $p = mv$, this relationship can be written as

$$E = \frac{p^2}{2m}.$$

Substituting our expressions for E and p, and simplifying, gives

$$\omega = \frac{hk^2}{4\pi m}$$

for the dispersion relation. The phase velocity v_p is $\omega/k = hk/4\pi m$, and since $p = hk/2\pi$ we have $v_p = p/2m = mv/2m = v/2$. The group velocity v_g is $d\omega/dk = hk/2\pi m = 2v_p = v$.

[The result for the phase velocity may seem surprising, but we should recall that it is the group velocity that determines how fast the associated particle travels and that this is therefore the one which we expect to be equal to v. We can show that the group velocity is equal to the particle velocity, even if the latter is relativistic, by using the relativistic mass–energy relation:

$$E^2 = p^2 c^2 + m_0^2 c^4.$$

Putting $E = h\omega/2\pi$ and $p = hk/2\pi$ gives

$$\frac{h^2 \omega^2}{4\pi^2} = \frac{h^2 c^2 k^2}{4\pi^2} + m_0^2 c^4$$

for the dispersion relation, and differentiating this with respect to k gives

$$\frac{d\omega}{dk} = \frac{c^2 k}{\omega} = \frac{c^2 p}{E}.$$

Since $p = \gamma m_0 v$ and $E = \gamma m_0 c^2$, where γ is the Lorentz factor, it follows that $d\omega/dk = v$.]

The de Broglie wave enters the slit from the left, i.e. from $z < 0$, as shown in figure 68. We know from Fraunhofer diffraction theory (see problem 113) that when a plane parallel wave of wavelength λ falls on a long, narrow slit of width a, the wave will be expanded into a 'fan' with maximum intensity in the z-direction and an angular width in the xz-plane of the order of λ/a (assuming that $\lambda \ll a$). Its width in the yz-plane will be negligible.

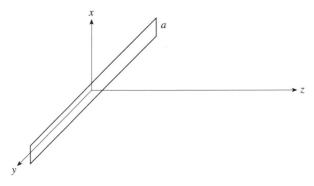

Figure 68

The interpretation of this phenomenon in terms of particles is as follows. At $z < 0$ a particle will have momentum $p_z = h/\lambda$ parallel to the z-direction, and $p_x = p_y = 0$. However, when the particle is within the slit it has a maximum uncertainty $\Delta x \approx a$ in its x-coordinate, though its y-coordinate has a large uncertainty. The component p_x of the momentum must therefore acquire a spread $\Delta p_x \approx h/a$, centred (by symmetry) on $p_x = 0$. The component p_y will be unaffected, since $\Delta p_y \approx 0$. Now a particle which has components of momentum $(p_x, 0, p_z)$ makes an angle $\tan^{-1}(p_x/p_z)$ with the z-axis, and if $p_x \ll p_z$ this angle is approximately p_x/p_z. Thus the spread of directions must be $\Delta p_x/p_z = (h/a) \div (h/\lambda) = \lambda/a$, as before. The two explanations (wave and particle) are therefore consistent.

Problem 78

Free neutrons have a decay constant of $1.10 \times 10^{-3}\ \mathrm{s}^{-1}$. If the de Broglie wavelength of the neutrons in a parallel beam is 1 nm, determine the distance from the source where the beam intensity has dropped to half its starting value.

Solution

The momentum p and the de Broglie wavelength are related by

$$p = h/\lambda,$$

so the neutron velocity v is given by

$$v = h/m\lambda$$

if the velocity is non-relativistic. Inserting values into this expression, we find that it gives $v \approx 400 \text{ m s}^{-1}$, so the assumption is very well justified.

If the decay constant is k, the number of neutrons remaining at time t is given by $N_0 \exp(-kt)$ where N_0 is the number at time zero, so the intensity of the beam will be given by $I_0 \exp(-kt)$ where I_0 is the intensity at the source. The time required for the intensity to fall to half of its starting value is thus given by $(\ln 2)/k$, and the distance travelled by the beam in this time is

$$\begin{aligned} x &= v(\ln 2)/k \\ &= h(\ln 2)/mk\lambda. \end{aligned}$$

Substituting the values of k and λ, and taking $m = 1.67 \times 10^{-27}$ kg, gives
▶ $x = 2.5 \times 10^5$ m.

Problem 79

Using Bohr theory, derive an expression for the frequencies at which light is absorbed by hydrogen atoms in their ground state. Why would these frequencies be slightly different for tritium atoms?

Solution

Let us consider an electron of mass m and a singly charged nucleus of mass M separated by a distance r and both describing circular orbits of angular frequency ω. The circular orbits of the two particles will be centred on the centre of mass of the system which is at a distance

$$\frac{mr}{M + m}$$

from the nucleus, as shown in figure 69.

The orbital speed of the electron is thus

$$\frac{Mr\omega}{M + m},$$

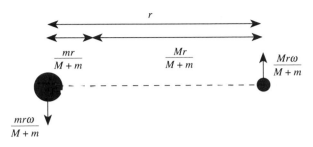

Figure 69

so its angular momentum about the centre of mass is

$$m\left(\frac{Mr\omega}{M+m}\right)\left(\frac{Mr}{M+m}\right) = \frac{M^2 m \omega r^2}{(M+m)^2}.$$

Similarly, the orbital speed of the nucleus is

$$\frac{mr\omega}{M+m},$$

so its angular momentum is

$$M\left(\frac{mr\omega}{M+m}\right)\left(\frac{mr}{M+m}\right) = \frac{Mm^2 \omega r^2}{(M+m)^2}.$$

The total angular momentum J of the system is thus

$$J = \frac{Mm\omega r^2}{(M+m)^2}(M+m) = \frac{Mm\omega r^2}{M+m}.$$

Now if we define the *reduced mass* of the system as

$$\mu = \frac{Mm}{M+m}, \tag{1}$$

we can write the angular momentum as

$$J = \mu\omega r^2. \tag{2}$$

We assume that the angular momentum is quantised in units of $h/2\pi$, where h is Planck's constant, so that equation (2) can be written as

$$\omega r^2 = \frac{nh}{2\pi\mu}. \tag{3}$$

[If $M = \infty$, this assumption is equivalent to requiring that the circumference of the electron's orbit round the nucleus contains an integral number of de Broglie wavelengths.]

The electrostatic force acting on the electron is

$$\frac{e^2}{4\pi\varepsilon_0 r^2}$$

and this must provide the centripetal force

$$m\omega^2\left(\frac{Mr}{M+m}\right) = \mu\omega^2 r.$$

Equating these, rearranging, gives

$$\omega^2 r^3 = \frac{e^2}{4\pi\varepsilon_0\mu}. \tag{4}$$

We now have two equations, (3) and (4), relating ω and r. We can therefore solve them to find ω and r. First, we square equation (3) and divide by equation (4) to obtain

$$r = \frac{\varepsilon_0 h^2 n^2}{\pi\mu e^2}. \tag{5}$$

Next, we substitute this into either (3) or (4) to obtain

$$\omega = \frac{\pi\mu e^4}{2\varepsilon_0^2 h^3 n^3}. \tag{6}$$

Now we can calculate the total energy of the system. The potential energy is just

$$E_{\mathrm{p}} = \frac{-e^2}{4\pi\varepsilon_0 r},$$

so substituting for r from equation (5) gives

$$E_{\mathrm{p}} = \frac{-\mu e^4}{4\varepsilon_0^2 h^2 n^2}.$$

The kinetic energy of the electron is

$$\frac{1}{2}m\omega^2\left(\frac{Mr}{M+m}\right)^2 = \frac{1}{2}\omega^2 r^2\frac{M^2 m}{(M+m)^2}$$

and the kinetic energy of the nucleus is

$$\frac{1}{2}M\omega^2\left(\frac{mr}{M+m}\right)^2 = \frac{1}{2}\omega^2 r^2\frac{Mm^2}{(M+m)^2},$$

so the total kinetic energy E_{k} is

$$E_{\mathrm{k}} = \frac{1}{2}\omega^2 r^2\frac{Mm}{M+m} = \frac{1}{2}\mu\omega^2 r^2.$$

Substituting for r from equation (5) and for ω from equation (6) gives

$$E_k = \frac{\mu e^4}{8\varepsilon_0^2 h^2 n^2}.$$

Finally, adding the expressions for the potential and kinetic energy gives the total energy E as

$$E = \frac{-\mu e^4}{8\varepsilon_0^2 h^2 n^2}. \tag{7}$$

The ground state of the atom is the lowest possible energy, which is given by $n = 1$. Thus the energy required to excite the atom from the ground state to a higher state is given by

$$\frac{\mu e^4}{8\varepsilon_0^2 h^2}\left(1 - \frac{1}{n^2}\right),$$

where n is an integer greater than 1. If this energy is supplied by a photon of frequency v, we can equate this expression to hv to obtain the frequency v of the transition:

$$v = \frac{\mu e^4}{8\varepsilon_0^2 h^3}\left(1 - \frac{1}{n^2}\right).$$

For hydrogen atoms, the nuclear mass M is 1836 times larger than the electron mass m, so the reduced mass μ is

$1836m/1837.$

For tritium atoms, the nuclear mass is about 5496 times larger than the electron mass, so μ is

$5496m/5497.$

Thus the frequencies for tritium will be higher than those for hydrogen by a factor of

$$\frac{5496}{5497} \times \frac{1837}{1836} = 1.000\,36.$$

Problem 80

Positronium is a hydrogen-like system with the proton replaced by a positron. Use the Bohr model to calculate the energy of the system in its ground state.

Solution

From the previous problem, we know that the ground state energy of a hydrogen-like atom is

$$E = - \frac{\mu e^4}{8\varepsilon_0^2 h^2},$$

where μ is the reduced mass of the electron, defined by $1/\mu = 1/m + 1/M$ where m is the mass of the electron and M is the mass of the nucleus.

For positronium, $M = m$ so $\mu = m/2$. Substituting this into the expression for the ground-state energy gives $E = -1.08 \times 10^{-18}$ J. Converting this to electron volts gives $E = -6.8$ eV.

[Note that this is half the value for the hydrogen atom.]

Problem 81

Calculate the velocity of an electron in the third Bohr orbit ($n = 3$) of the hydrogen atom.

Solution

From problem 79, we have

$$v = \frac{Mr\omega}{M + m},$$

where

$$r = \frac{\varepsilon_0 h^2 n^2}{\pi \mu e^2}$$

and

$$\omega = \frac{\pi \mu e^4}{2\varepsilon_0^2 h^3 n^3}.$$

Substituting these expressions for r and ω into the expression for v gives

$$v = \frac{Me^2}{2(M + m)\varepsilon_0 hn}.$$

Taking $M + m \approx M$, and setting $n = 3$, gives $v = 7.3 \times 10^5$ m s^{-1}.

Problem 82

An atom has energy levels $E_n = -A/n^2$ where n is an integer and A is a constant. Among the spectral lines that the atom can absorb at room temperature are two adjacent lines with wavelengths 97.5 nm and 102.8 nm. Find the value of the constant A in electron volts.

Solution

An absorption line corresponds to the atom being excited from a lower to a higher energy level, i.e. from quantum number n_1 to quantum number n_2 where $n_2 > n_1$. The energy required to effect this transition is clearly

$$E(n_1 \to n_2) = \frac{A}{n_1^2} - \frac{A}{n_2^2},$$

so the wavelength of the absorption line must be

$$\lambda(n_1 \to n_2) = \frac{ch}{A}\left(\frac{1}{n_1^2} - \frac{1}{n_2^2}\right)^{-1},$$

where c is the speed of light and h is Planck's constant. Adjacent absorption lines must correspond to transitions $n_1 \to n_2$ and $n_1 \to n_2 + 1$. The latter transition requires a larger energy so the wavelength will be smaller. Since we do not yet know the value of A, we must first identify the values of n_1 and n_2 corresponding to the two observed wavelengths, and we can do this by taking the ratio of the two wavelengths to eliminate the dependence on A:

$$\frac{\lambda(n_1 \to n_2)}{\lambda(n_1 \to n_2 + 1)} = \frac{\dfrac{1}{n_1^2} - \dfrac{1}{(n_2 + 1)^2}}{\dfrac{1}{n_1^2} - \dfrac{1}{n_2^2}} = \frac{102.8}{97.5} = 1.054.$$

To solve this, we can rearrange it to give

$$\frac{0.054}{n_1^2} = \frac{1.054}{n_2^2} - \frac{1}{(n_2 + 1)^2}.$$

Taking $n_2 = 2, 3, 4, \ldots$, we can use this expression to calculate the corresponding value of n_1. As soon as we obtain a value which is close enough to an integer we can assume that we have found the appropriate values of n_1 and n_2:

$n_2 = 2$ gives $n_1 = 0.595$,
$n_2 = 3$ gives $n_1 = 0.994$,
$n_2 = 4$ gives $n_1 = 1.445$, etc.

So it is clear that $n_1 = 1$ and $n_2 = 3$ will give the observed ratio of wavelengths. We can now substitute these values into our expression for $\lambda(n_1 \rightarrow n_2)$ to obtain a value for A. Using the larger wavelength we obtain

$$A = \frac{2.998 \times 10^8 \times 6.626 \times 10^{-34}}{102.8 \times 10^{-9}} \left(\frac{1}{1} - \frac{1}{9}\right)^{-1} \text{J} = 2.174 \times 10^{-18} \text{ J}$$

and using the smaller wavelength we obtain

$$A = \frac{2.998 \times 10^8 \times 6.626 \times 10^{-34}}{97.5 \times 10^{-9}} \left(\frac{1}{1} - \frac{1}{16}\right)^{-1} \text{J} = 2.173 \times 10^{-18} \text{ J.}$$

▶ Converting these to electron volts using the fact that the electron charge is 1.602×10^{-19} C, and averaging, gives a value for A of 13.6 eV to the precision warranted by the data. [We might recall that the ionisation potential of hydrogen is 13.60 V, so the atom in question must be hydrogen.]

Problem 83

A spectrometer of resolving power 5×10^5 is used to observe the Balmer series in the hydrogen spectrum (transitions for which the principal quantum number changes from $n > 2$ to $n = 2$). Find the quantum number of the highest level for which the line would just be resolved from its neighbours. Neglect all sources of line width in the light source.

Solution

The wavelength of a line in the Balmer series is given by

$$\frac{1}{\lambda} = K\left(\frac{1}{4} - \frac{1}{n^2}\right),$$

where K is a constant and n is the quantum number of the higher level. [As we shall see, we do not need to know the value of K.] We need to know the change in λ when n changes by 1, i.e. $\Delta\lambda$ corresponding to $\Delta n = 1$. If the fractional changes in λ and n are small enough (i.e. n is large enough), $\Delta\lambda/\Delta n \approx d\lambda/dn$, so we will differentiate the expression for $1/\lambda$:

$$-\frac{1}{\lambda^2}\frac{d\lambda}{dn} = \frac{2K}{n^3};$$

thus

$$\frac{d\lambda}{dn} = -\frac{2K\lambda^2}{n^3}.$$

The fractional wavelength change $\Delta\lambda/\lambda$ corresponding to a change $\Delta n = 1$ is therefore given approximately by

$$\frac{\Delta\lambda}{\lambda} \approx \frac{2K\lambda}{n^3} = \frac{2}{n^3\left(\dfrac{1}{4} - \dfrac{1}{n^2}\right)}$$

(we have dropped the minus sign since we are only interested in the magnitude of the change in λ). Now the resolving power R of a spectrometer is defined as the largest value of $\lambda/\Delta\lambda$ which it can resolve (see problem 115), so we have

$$\frac{n^3}{4} - n < 2R.$$

Taking $R = 5 \times 10^5$ and ignoring the term n in relation to the term $n^3/4$ gives $n^3 < 4 \times 10^6$ and hence $n < 158.7$. Thus the spectrometer should be able to resolve lines up to about $n = 158$.

▶

Problem 84

State the Schrödinger equation for a particle of mass m moving in the xy-plane subject to a potential energy $V(x, y)$. What is the probability of finding it in a small area ΔS centred at the point (x, y) where the wavefunction is $\psi(x, y)$?

A particle of mass m is confined to a line and has a wavefunction $\psi = C\exp(-\alpha^2 x^2/2)$. Calculate C in terms of α and obtain an expression for the potential energy at distance x from the origin if the total energy is $h^2\alpha^2/8\pi^2 m$. Write down an integral expression for the probability of finding the particle between the points $x = 4$ and $x = 5$.

(You may assume that

$$\int_{-\infty}^{\infty} \exp(-y^2)\, dy = \sqrt{\pi}.)$$

Solution

Schrödinger's (time-independent) equation is, in two dimensions,

$$\frac{h^2}{8\pi^2 m}\left(\frac{\partial^2 \psi}{\partial x^2} + \frac{\partial^2 \psi}{\partial y^2}\right) + (E - V)\psi = 0,$$

where ψ is the wavefunction. The probability of finding the particle in an area ΔS is given by $\psi^*\psi\Delta S$.

If $\psi = C\exp(-\alpha^2 x^2/2)$, the probability density $p(x) = \psi^*\psi$ is $C^2\exp(-\alpha^2 x^2)$. This probability density must be normalised so that

$$\int_{-\infty}^{\infty} p(x)\,dx = 1;$$

thus

$$C^2\int_{-\infty}^{\infty} \exp(-\alpha^2 x^2)\,dx = \frac{C^2}{\alpha}\int_{-\infty}^{\infty} \exp(-y^2)\,dy = \frac{C^2\sqrt{\pi}}{\alpha} = 1.$$

So

$$C = \alpha^{1/2}\pi^{-1/4}.$$

If $\psi = C\exp(-\alpha^2 x^2/2)$,

$$d\psi/dx = -\alpha^2 xC\exp(-\alpha^2 x^2/2)$$

and

$$d^2\psi/dx^2 = -\alpha^2 C\exp(-\alpha^2 x^2/2) + \alpha^4 x^2 C\exp(-\alpha^2 x^2/2)$$
$$= \alpha^2\psi(\alpha^2 x^2 - 1).$$

Substituting into Schrödinger's equation, and including the information that $E = h^2\alpha^2/8\pi^2 m$, thus gives

$$\frac{h^2\alpha^2}{8\pi^2 m}(\alpha^2 x^2 - 1) + \frac{h^2\alpha^2}{8\pi^2 m} - V = 0,$$

so $V = h^2\alpha^4 x^2/8\pi^2 m$.

[We recognise this as the potential of a simple harmonic oscillator, since the classical restoring force $-dV/dx$ is proportional to $-x$. The constant of proportionality is the classical spring constant k, which thus has a value of $h^2\alpha^4/4\pi^2 m$. The angular frequency ω of a mass m on a spring of spring constant k is given by $\omega^2 = k/m$, so in this case, $\omega = h\alpha^2/2\pi m$. Comparing this with the quoted energy of the quantum state, we see that $E = h\omega/4\pi = h\nu/2$. This is in fact the *zero-point energy* of a one-dimensional harmonic oscillator.]

The probability of finding the particle between $x = 4$ and $x = 5$ is given by the integral of the probability density:

$$p(4, 5) = \int_4^5 \psi^*\psi\,dx = C^2\int_4^5 \exp(-\alpha^2 x^2)\,dx = \frac{\alpha}{\sqrt{\pi}}\int_4^5 \exp(-\alpha^2 x^2)\,dx.$$

[We could evaluate this in terms of the error function.]

Problem 85

Calculate the possible values of the energy of a particle of mass m situated inside a deep one-dimensional potential well of width a described by $V = 0$ for $0 \leqslant x \leqslant a$ and $V = \infty$ elsewhere.

An electron is confined within a thin layer of a semiconductor. If the layer can be treated as a deep one-dimensional well, calculate its thickness if the difference in energy between the first and second levels is 0.05 eV.

Solution

The Schrödinger equation in one dimension is

$$-\frac{h^2}{8\pi^2 m}\frac{d^2\psi}{dx^2} + V\psi = E\psi.$$

When $V = 0$ (inside the well), this becomes

$$E\psi = -\frac{h^2}{8\pi^2 m}\frac{d^2\psi}{dx^2}.$$

We recognise this as a differential equation having sinusoidal or cosinusoidal solutions, and try

$$\psi = A\cos(kx) + B\sin(kx).$$

With this form of ψ, $d^2\psi/dx^2 = -k^2\psi$, so the energy is

$$E = \frac{h^2 k^2}{8\pi^2 m}.$$

Since the potential V is infinite for $x < 0$ and for $x > a$, the wavefunction ψ must be zero at these values of x. The condition that $\psi(0) = 0$ clearly requires that $A = 0$, and the condition that $\psi(a) = 0$ requires that

$$ka = \pi n,$$

where n is a positive integer. $n = 0$ is not allowed, since it would give $\psi = 0$ everywhere. Figure 70 shows the first three solutions (see problem 103).

Substituting these values of k into our expression for E gives

$$E = \frac{n^2 h^2}{8ma^2}.$$

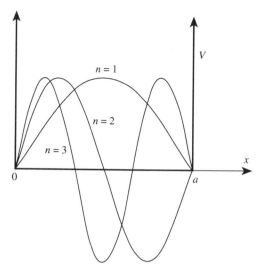

Figure 70

The lowest energy is thus $h^2/8ma^2$, and the second level will have an energy $4h^2/8ma^2$, so the difference between the first and second levels must be $3h^2/8ma^2$. Equating this to 0.05 eV $= 8.0 \times 10^{-21}$ J, and substituting the electron mass for m, gives $a = 4.8$ nm.

Problem 86

Consider particles of mass m and energy E moving along the x-axis under the influence of a potential $V(x)$. The particles have energy $E > 0$ and the potential has the form

$$V(x) = 0 \text{ for } x < 0,$$
$$V(x) = V_0 \text{ for } x \geqslant 0,$$

where V_0 is a constant such that $V_0 > E$.

(i) For $x < 0$ show that there are solutions of the Schrödinger equation of the form $\psi(x) = \exp(ikx)$. Determine the two possible values of k and use the appropriate operator to determine the corresponding values of the momentum.

(ii) A beam of particles travels in the positive x-direction from negative x. Describe qualitatively what happens to the particles in the beam according to classical mechanics and according to quantum mechanics.

(iii) For positive x consider electrons with $E = 1$ eV and the potential with $V_0 = 2$ eV. If $\psi(x)$ is the wavefunction at the point x, where x is measured in units of 10^{-10} m, determine $|\psi(1)/\psi(0)|^2$. What does this quantity represent?

Solution

Figure 71 shows the potential $V(x)$ and the total energy E.

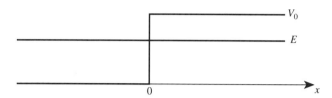

Figure 71

(i) In one dimension, the Schrödinger equation can be written

$$\frac{h^2}{8\pi^2 m}\frac{d^2\psi}{dx^2} + (E - V)\psi = 0$$

and in the region $x < 0$ we have $V = 0$ so the equation becomes

$$\frac{h^2}{8\pi^2 m}\frac{d^2\psi}{dx^2} + E\psi = 0.$$

Let us try a solution of the form $\psi = \exp(ikx)$. $d^2\psi/dx^2$ is $-k^2 \exp(ikx)$ which is equal to $-k^2\psi$, so the Schrödinger equation requires that

$$-\frac{h^2 k^2}{8\pi^2 m}\psi + E\psi = 0.$$

This can clearly be satisfied (so that the assumed form of ψ is a possible solution of the Schrödinger equation) if

$$k^2 = \frac{8\pi^2 Em}{h^2} \quad \therefore k = \pm\sqrt{\frac{8\pi^2 Em}{h^2}}.$$

The momentum operator is

$$\frac{h}{2\pi i}\frac{\partial}{\partial x}$$

such that the momentum p of a particle with a one-dimensional wavefunction ψ satisfies the equation

$$\frac{h}{2\pi i}\frac{d\psi}{dx} = p\psi.$$

If $\psi = \exp(ikx)$, $d\psi/dx = ik\psi$ so that $p\psi = hk\psi/2\pi$. The momentum p is thus given by $hk/2\pi$, so the two solutions correspond to momenta of

$$p = \pm \frac{h}{2\pi} \sqrt{\frac{8\pi^2 Em}{h^2}} = \pm \sqrt{(2Em)}.$$

[These are identical to the equivalent classical expressions, since, classically, $E = mv^2/2$ so that $\sqrt{(2Em)} = mv$, which is the classical momentum.]

(ii) According to classical mechanics, the particles would have negative kinetic energy in the region $x \geq 0$. Since this is impossible, the particles would be reflected (elastically scattered) at $x = 0$. Quantum-mechanically, the particles are also reflected, but not from a definite value of x. The wavefunction ψ does not fall abruptly to zero at $x = 0$ but instead diminishes exponentially with increasingly positive values of x (it is an *evanescent wave*), so that the probability of finding a particle at $x > 0$ decreases exponentially with increasing x.

(iii) To find the ratio of ψ at $x = 10^{-10}$ m to its value at $x = 0$, we need to know the form of $\psi(x)$ for $x \geq 0$. Returning to the Schrödinger equation, we can write

$$\frac{h^2}{8\pi^2 m} \frac{d^2\psi}{dx^2} + T\psi = 0,$$

where $T = E - V_0$ is the classical kinetic energy, which (as we noted before) is negative. By analogy with the solution $\psi = \exp(ikx)$, we can write the solutions to this differential equation as $\psi = \exp(i\kappa x)$, where

$$\kappa = \pm \sqrt{\frac{8\pi^2 Tm}{h^2}}.$$

Since T is negative, these values of κ are imaginary, so we can write them as

$$\kappa = \pm i\sqrt{\frac{8\pi^2 m(V_0 - E)}{h^2}} = \pm i\alpha,$$

where α is real. Substituting these values of κ into $\psi = \exp(i\kappa x)$ gives the two solutions $\psi = \exp(\alpha x)$ and $\psi = \exp(-\alpha x)$, and we can reject the solution $\psi = \exp(\alpha x)$ since it becomes infinitely large as x tends to infinity. Finally, then, we can write the general solution when $x \geq 0$ as $\psi = A \exp(-\alpha x)$, where A is a constant. Taking $V_0 - E = 1$ eV $= 1.60 \times 10^{-19}$ J and $m = 9.11 \times 10^{-31}$ kg gives $\alpha = 5.12 \times 10^9$ m^{-1}, so over a distance of 10^{-10} m, $|\psi|$ falls by a factor of $\exp 0.512 = 1.67$. The value of $|\psi(1)/\psi(0)|^2$ is thus $1.67^{-2} = 0.36$. This is the ratio of the probability of finding an electron at $x = 10^{-10}$ m to that of finding one at $x = 0$.

Problem 87

^{238}U decays with a half-life of 4.51×10^9 years, the decay series eventually ending at ^{206}Pb, which is stable. A rock sample analysis shows that the ratio of the numbers of atoms of ^{206}Pb to ^{238}U is 0.0058. Assuming that all the ^{206}Pb has been produced by the decay of ^{238}U and that all other half-lives in the chain are negligible, calculate the age of the rock sample.

Solution

After a time t, a sample of ^{238}U originally consisting of N atoms will have decayed to $N \exp(-\lambda t)$ atoms, where the decay constant λ is given by

$$\lambda = \frac{\ln 2}{t_{1/2}}$$

and $t_{1/2}$ is the half-life. We are told (in effect) that we may assume that all the decayed atoms have become ^{206}Pb, so the number of these atoms is $N(1 - \exp[-\lambda t])$. The ratio of lead to uranium is thus

$$\frac{1 - \exp(-\lambda t)}{\exp(-\lambda t)} = \exp(\lambda t) - 1.$$

▶ Taking this ratio as 0.0058 we have $\lambda t = \ln 1.0058 = 0.005\,78$, and taking λ as $(\ln 2)/(4.51 \times 10^9 \text{ years}) = 1.54 \times 10^{-10} \text{ year}^{-1}$ gives $t = 38 \times 10^6$ years.

Problem 88

A certain uranium ore contains both $^{235}_{92}$U and $^{238}_{92}$U. Analysis shows that it contains 0.80 g of $^{206}_{82}$Pb for each gram of the relevant uranium isotope.
  (a) Determine the age of the ore in years.
  (b) If the sample initially contained 3.00 mg of $^{235}_{92}$U, how much remains now?
  (c) Determine the present-day activity due to the $^{235}_{92}$U.
$(\lambda(^{235}\text{U}) = 3.08 \times 10^{-17} \text{ s}^{-1}; \lambda(^{238}\text{U}) = 4.87 \times 10^{-18} \text{ s}^{-1}.)$

Solution

(a) The first thing we need to do is to decide whether $^{206}_{82}$Pb results from the radioactive decay of $^{235}_{92}$U or of $^{238}_{92}$U. α-emission causes the atomic number Z to decrease by 2 and the nucleon number A to decrease by 4. β^--emission causes Z to increase by 1, but does not change A. β^+-emission causes Z to decrease by 1 and also leaves A unchanged, and γ-emission alters neither Z nor A. We can see from this that, during a decay series, the nucleon number A must change by integral multiples of 4 at each step. This observation identifies $^{238}_{92}U$ as the originator of the series terminating at $^{206}_{82}Pb$, since their values of A differ by 32 which is a multiple of 4. $^{235}_{92}U$ cannot be a member of the series since its nucleon number differs from that of $^{206}_{82}Pb$ by 29, which is not a multiple of 4.

Let us suppose that at the time the ore was formed there were N atoms of ^{238}U. After time t there will be $N\exp(-\lambda t)$ atoms of ^{238}U, where λ is the appropriate decay constant, and $N(1-\exp[-\lambda t])$ atoms of ^{206}Pb. The ratio of the mass of ^{206}Pb to the mass of ^{238}U is thus

$$\frac{m_{Pb}(1-\exp[-\lambda t])}{m_U \exp(-\lambda t)} = \frac{206(1-\exp[-\lambda t])}{238\exp(-\lambda t)} = 0.80,$$

where m_{Pb} is the mass of an atom of the lead isotope, and similarly for the uranium isotope.

Rearranging this expression, we find

$$\exp(\lambda t) - 1 = \frac{0.80 \times 238}{206} = 0.92,$$

so that $\lambda t = 0.65$. Taking $\lambda = 4.87 \times 10^{-18}\,s^{-1}$ gives $t = 1.34 \times 10^{17}s = 4.3 \times 10^9$ years. [This is approximately the estimated age of the Earth.]

(b) In the same period of time, the amount of ^{235}U will have decayed to a fraction $\exp(-1.34 \times 10^{17} \times 3.08 \times 10^{-17}) = \exp(-4.13) = 0.0161$ of its original amount, so there will be $3.00 \times 0.0161\,mg = 48\,\mu g$.

(c) The activity in becquerels is the number of disintegrations per second. If there are N undecayed atoms, the activity is λN. Since the mass of ^{235}U is 48×10^{-9} kg and the mass of an atom of ^{235}U is approximately $235 \times 1.66 \times 10^{-27}$ kg $= 3.90 \times 10^{-25}$ kg, $N = 1.23 \times 10^{17}$. Taking $\lambda = 3.08 \times 10^{-17}\,s^{-1}$ gives the activity as 3.8 Bq.

Problem 89

The following deuterium reactions and corresponding reaction energies Q are found to occur.

$^{14}N(d, p)^{15}N$, $Q = 8.53\,MeV$,
$^{15}N(d, \alpha)^{13}C$, $Q = 7.58\,MeV$,
$^{13}C(d, \alpha)^{11}B$, $Q = 5.16\,MeV$.

What is the Q value of the reaction $^{11}B(\alpha, n)^{14}N$?

(The notation $^{14}N(d, p)^{15}N$ represents the reaction $^{14}N + d \rightarrow ^{15}N + p$. $^{4}_{2}He = 4.0026m_u$, $^{2}_{1}H = 2.0140m_u$, $^{1}_{1}H = 1.0078m_u$, n $= 1.0087m_u$. One twelfth of the mass of $^{12}C = 931$ MeV/c^2.)

Solution

The Q value of the first reaction implies that

$$^{14}N + d = ^{15}N + p + 8.53 \text{ MeV},$$

where ^{14}N represents the mass of the ^{14}N nucleus in energy units, and so on. This can be rearranged to give

$$^{14}N - ^{15}N = p - d + 8.53 \text{ MeV}.$$

The second reaction similarly implies that

$$^{15}N - ^{13}C = \alpha - d + 7.58 \text{ MeV},$$

and the third reaction implies that

$$^{13}C - ^{11}B = \alpha - d + 5.16 \text{ MeV}.$$

Adding these three equations gives

$$^{14}N - ^{11}B = p + 2\alpha - 3d + 21.27 \text{ MeV}.$$

If we write Q for the reaction energy of $^{11}B + \alpha \rightarrow ^{14}N + n$, we must have

$$Q = ^{11}B - ^{14}N + \alpha - n = 3d - \alpha - p - n - 21.27 \text{ MeV}.$$

Now

$$3d - \alpha - p - n = (3 \times 2.0140 - 4.0026 - 1.0078 - 1.0087)m_u$$
$$= 0.0229m_u$$

and

$$1m_u = 931 \text{ MeV}$$

so

▶ $$Q = 0.0229 \times 931 - 21.27 \text{ MeV} = 0.05 \text{ MeV}.$$

Problem 90

It is proposed to use $^{113}_{48}Cd$ as an attenuator in a nuclear pile. Calculate the thickness required to attenuate a neutron beam to 0.01% of its original intensity if it has a density of 8.7×10^3 kg m^{-3} and a cross section of 2.1×10^4 barns. (1 barn $= 10^{-28}$ m^2.)

Solution

Consider a slab of the attenuator with area A and thickness dz. If the number of atoms per unit volume is n, the slab contains $nA\,dz$ atoms, and if we write σ for the cross-section of each atom, the total cross section of the slab is $\sigma n A\,dz$. This is a fraction $\sigma n\,dz$ of the slab's total area, so the intensity of the beam is reduced by a fraction $\sigma n\,dz$ on passing through the slab. We can thus write the differential equation describing the intensity I of the beam as

$$\frac{dI}{I} = -\sigma n\,dz,$$

which has the solution

$$I = I_0 \exp(-\sigma n z).$$

If the intensity is to be reduced to $0.01\% = 10^{-4}$ of its original value, $\sigma n z = \ln 10^4 = 9.21$.

The number density n of atoms is given by ρ/m where ρ is the density and m is the atomic mass. ρ is given as 8.7×10^3 kg m^{-3} and we may take m as $113 \times 1.66 \times 10^{-27}$ kg $= 1.88 \times 10^{-25}$ kg, giving $n = 4.63 \times 10^{28}$ m^{-3}. Thus

$$z = \frac{9.21}{2.1 \times 10^4 \times 10^{-28} \times 4.63 \times 10^{28}}\ \text{m} = 9.5 \times 10^{-5}\ \text{m}.$$

▶ The thickness required is thus 95 μm.

Oscillations and waves

Show that the periods of a pendulum bob are the same for linear oscillation and circular motion (for small displacements).

Solution

Consider a pendulum of length l describing linear oscillations (i.e. oscillations in a plane), as shown in figure 72.

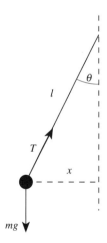

Figure 72

We will assume that the horizontal displacement x of the pendulum bob is small enough that the vertical component of the bob's motion can be ignored. This requires that there is no net vertical force on the bob. If T is the tension in the string and mg is the weight of the pendulum bob, resolving vertically gives

$$T \cos \theta = mg$$

and resolving horizontally gives

$$T \sin \theta = -m\, d^2x/dt^2,$$

166

so by taking the ratio of these two expressions we obtain

$$d^2x/dt^2 = -g\tan\theta.$$

Now for small values of θ, $\tan\theta \approx x/l$, so we can write the equation of motion as

$$d^2x/dt^2 = -gx/l.$$

The solution of this differential equation is simple harmonic motion of x, with angular frequency

▶ $$\omega = (g/l)^{1/2}.$$

[This result can also be obtained by considering the energy of the system. In terms of the angle θ, the potential energy of the system is

$$mgl(1 - \cos\theta)$$

plus an arbitrary constant, and the kinetic energy is

$$\frac{1}{2}ml^2\left(\frac{d\theta}{dt}\right)^2.$$

The total energy E is constant, so the rate of change of E with time must be zero. Thus

$$\frac{dE}{dt} = mgl\sin\theta\frac{d\theta}{dt} + ml^2\frac{d\theta}{dt}\frac{d^2\theta}{dt^2} = 0$$

and hence

$$\frac{d^2\theta}{dt^2} = -\frac{g\sin\theta}{l}.$$

If we put $\sin\theta \approx \theta$, we obtain the differential equation of simple harmonic motion with angular frequency $\omega = (g/l)^{1/2}$, as before.]

Now consider circular motion of the bob. The same diagram can be used, but now the term d^2x/dt^2 is the centripetal acceleration of the bob, so if Ω is the angular frequency of rotation,

$$gx/l = \Omega^2 x$$

and hence

▶ $$\Omega = (g/l)^{1/2} = \omega$$

as required.

Problem 92

Two bodies of masses m and $3m$ are attached to each other and to two fixed points by three identical light springs as shown in figure 73.

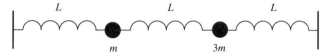

Figure 73

The whole arrangement rests on a smooth horizontal table. The springs are stretched so that the tension in each spring is T and its length is L (much greater than its unstretched length). Show that the angular frequencies of the normal modes for longitudinal oscillations of small amplitude are given by

$$\omega^2 = \frac{4 \pm \sqrt{7}}{3} \frac{T}{mL}.$$

Describe the motions of the two bodies for each normal mode.

Solution

Let us call the spring constants of the springs k, and consider the forces acting when the body of mass m (call this body 1) is displaced from its equilibrium position by x_1 and the body of mass $3m$ (body 2) is displaced by x_2 (see figure 74).

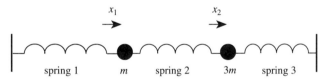

Figure 74

It is clear that spring 1 is extended by x_1 relative to its length at equilibrium, so the increase in its tension is kx_1. Similarly, spring 2 is extended by $x_2 - x_1$ relative to its length at equilibrium, so the increase in its tension is $k(x_2 - x_1)$. Thus the net force acting on body 1 must be kx_1 to the left from spring 1 and $k(x_2 - x_1)$ to the right from spring 2. The force in the direction of increasing x_1 is thus

$$k(x_2 - x_1) - kx_1 = -2kx_1 + kx_2.$$

By similar reasoning, the net force acting on body 2 is

$$kx_1 - 2kx_2,$$

so we can write the equations of motion of the two bodies as

$$m\, d^2x_1/dt^2 = -2kx_1 + kx_2, \tag{1}$$
$$3m\, d^2x_2/dt^2 = kx_1 - 2kx_2. \tag{2}$$

We assume that sinusoidal solutions exist, so we look for solutions of the form

$$x_1 = A\exp(i\omega t),$$
$$x_2 = B\exp(i\omega t),$$

where A and B are, in general, complex to take account of the possibility that the two bodies might oscillate out of phase with one another. Substituting these solutions into (1) and (2) gives

$$-m\omega^2 A = -2kA + kB \tag{3}$$

and

$$-3m\omega^2 B = kA - 2kB. \tag{4}$$

Rearranging (3) gives

$$(2k - m\omega^2)A = kB \tag{5}$$

and rearranging (4) gives

$$(2k - 3m\omega^2)B = kA. \tag{6}$$

Equations (5) and (6) can be combined by multiplying them together to eliminate A and B:

$$(2k - m\omega^2)(2k - 3m\omega^2) = k^2,$$

which can be rearranged to give

$$3m^2\omega^4 - 8km\omega^2 + 3k^2 = 0.$$

This is a quadratic equation in ω^2 which can be solved to give

$$\omega^2 = \frac{8km \pm \sqrt{(64k^2m^2 - 36k^2m^2)}}{6m^2} = \frac{4 \pm \sqrt{7}}{3}\frac{k}{m}.$$

[An alternative method of obtaining this result is to use matrix notation. Equations (3) and (4) can be rewritten as

$$-\frac{3m\omega^2 A}{k} = -6A + 3B,$$

$$-\frac{3m\omega^2 B}{k} = A - 2B,$$

which can be written in matrix form as

$$-\lambda \begin{bmatrix} A \\ B \end{bmatrix} = \begin{bmatrix} -6 & 3 \\ 1 & -2 \end{bmatrix} \begin{bmatrix} A \\ B \end{bmatrix},$$

where $\lambda = 3m\omega^2/k$ is an eigenvalue of the matrix. In order for this equation to be satisfied, the determinant of the matrix $\mathbf{M} + \lambda\mathbf{I}$ must be equal to zero, where $\mathbf{I}$ is the identity matrix and $\mathbf{M}$ is the matrix

$$\begin{bmatrix} -6 & 3 \\ 1 & -2 \end{bmatrix}.$$

Thus

$$\det \begin{bmatrix} -6 + \lambda & 3 \\ 1 & -2 + \lambda \end{bmatrix} = 0,$$

so

$$(-6 + \lambda)(-2 + \lambda) - 3 = 0,$$

which can be rearranged as a quadratic in λ:

$$\lambda^2 - 8\lambda + 9 = 0.$$

The solutions of this quadratic are

$$\lambda = 4 \pm \sqrt{7},$$

and recalling that $\lambda = 3m\omega^2/k$ we obtain

$$\omega^2 = \frac{4 \pm \sqrt{7}}{3} \frac{k}{m}$$

as before.]

Now if L is much greater than the unstretched lengths of the springs, the spring constant $k \approx T/L$, so finally we obtain

▶
$$\omega^2 = \frac{4 \pm \sqrt{7}}{3} \frac{T}{mL}$$

as required.

Our result for ω^2 can be substituted into equation (5) or equation (6) to find the ratio A/B. Choosing equation (6) for simplicity, we obtain

$$A/B = 2 - 3m\omega^2/k$$
$$= 2 - (4 \pm \sqrt{7}).$$

Thus the lower-frequency solution has $A/B = \sqrt{7} - 2$, which is real and positive, with a numerical value of 0.646. The two bodies thus oscillate in phase, the lighter body having an amplitude of oscillation 0.646 times that of the heavier body. Figure 75 shows the two extreme positions of this mode.

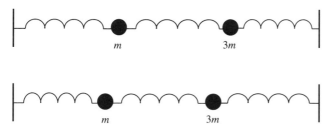

Figure 75

The higher-frequency solution has $A/B = -2 - \sqrt{7}$, which is real and negative, with a numerical value of -4.646. The two bodies oscillate in antiphase, the lighter body having an amplitude of oscillation 4.646 times that of the heavier body. Figure 76 shows the two extreme positions of this mode.

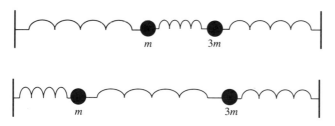

Figure 76

Problem 93

Two bodies of mass M_1 and M_2 are connected by a spring but are otherwise free to move along a horizontal line. A periodic force $F \cos \omega t$ is exerted on the body of mass M_1 along the line. Show that its displacement is given by

$$x_1 = F \cos \omega t \frac{k - \omega^2 M_2}{\omega^2 (\omega^2 M_1 M_2 - k[M_1 + M_2])},$$

where k is the spring constant (restoring force per unit extension). Indicate by a sketch graph the dependence of amplitude on frequency ω.

Solution

Write x_1 and x_2 for the respective displacements of the masses M_1 and M_2 from their equilibrium positions, as shown in figure 77. The extension of the spring is clearly $x_2 - x_1$, so the total force acting on M_1 is

$$k(x_2 - x_1) + F\exp(i\omega t) = M_1\, d^2x_1/dt^2 \tag{1}$$

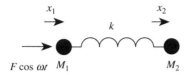

x_1 x_2

k

$F\cos\omega t$ M_1 M_2

Figure 77

and the total force acting on M_2 is

$$-k(x_2 - x_1) = M_2\, d^2x_2/dt^2 \tag{2}$$

(we have used the complex exponential notation to simplify the calculation).

Since we are forcing the system at frequency ω, let us try as a solution

$$x_1 = A\exp(i\omega t),$$
$$x_2 = B\exp(i\omega t).$$

Differentiating these twice with respect to time, and substituting into (1) and (2), gives

$$kB\exp(i\omega t) - kA\exp(i\omega t) + F\exp(i\omega t) = -M_1 A\omega^2\exp(i\omega t), \tag{1'}$$
$$-kB\exp(i\omega t) + kA\exp(i\omega t) = -M_2 B\omega^2\exp(i\omega t). \tag{2'}$$

Rearranging (2') to obtain B in terms of A gives

$$B = \frac{kA}{k - M_2\omega^2}.$$

Substituting this into (1') and eliminating $\exp(i\omega t)$ gives

$$\frac{k^2 A}{k - M_2\omega^2} - kA + F = -M_1 A\omega^2,$$

which can be rearranged to give

$$A = \frac{F(k - M_2\omega^2)}{\omega^2(\omega^2 M_1 M_2 - k[M_1 + M_2])}.$$

Thus

▶ $$x_1 = F \cos \omega t \; \frac{k - \omega^2 M_2}{\omega^2(\omega^2 M_1 M_2 - k[M_1 + M_2])}$$

as required.

The amplitude of the motion of M_1 is clearly the term A. We can see by inspection that it will have a value of zero when $\omega^2 = k/M_2$ and that it will be infinite (resonance) when

$$\omega^2 = k(M_1 + M_2)/(M_1 M_2) = k(1/M_1 + 1/M_2).$$

The amplitude will also be infinite when $\omega = 0$ (this just corresponds to a steady force accelerating the whole system), and it will tend to zero as ω tends to infinity.

Figure 78 shows a sketch of A as a function of ω.

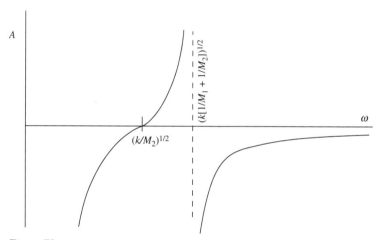

Figure 78

Problem 94

Derive the equation of motion of a particle of mass m subject to restoring and frictional forces of magnitude kx and $b\, dx/dt$ respectively, where x is its displacement and k and b are positive constants.

Show that $x = A \exp(-\gamma t) \cos(\omega t + \phi)$ is only a solution of the equation of motion for $4km > b^2$ and determine the value of γ. ω, γ, A and ϕ are real constants. Comment on the physical meaning of this solution.

An object oscillates harmonically with a frequency of 0.5 Hz and its amplitude of vibration is halved in 2 s. Find a differential equation for the oscillation.

Solution

Since the restoring force acts in the direction opposite to the displacement of the particle, it is given by the expression $-kx$. Similarly, the frictional force is $-b\,dx/dt$, and since the total force acting on the particle must equal the rate of change of its momentum, we may write the equation of motion as

$$m\,d^2x/dt^2 = -kx - b\,dx/dt$$

or

▶
$$m\,d^2x/dt^2 + b\,dx/dt + kx = 0.$$

Let us try $x = A\exp(-\gamma t)\cos(\omega t + \phi)$ as a solution of this equation. Repeated differentiation with respect to time gives

$$dx/dt = -A\gamma\exp(-\gamma t)\cos(\omega t + \phi) - A\omega\exp(-\gamma t)\sin(\omega t + \phi),$$
$$d^2x/dt^2 = A(\gamma^2 - \omega^2)\exp(-\gamma t)\cos(\omega t + \phi)$$
$$+ 2A\gamma\omega\exp(-\gamma t)\sin(\omega t + \phi).$$

Substituting these into the differential equation, and eliminating the factor $A\exp(-\gamma t)$, gives

$$(m\gamma^2 - m\omega^2 - b\gamma + k)\cos(\omega t + \phi) + (2m\gamma\omega - b\omega)\sin(\omega t + \phi) = 0.$$

Clearly, if this expression is to be true at all times, both $m\gamma^2 - m\omega^2 - b\gamma + k$ and $2m\gamma\omega - b\omega$ must be equal to zero. The second of these conditions implies that

▶
$$\gamma = b/2m.$$

Substituting this into the first condition, and rearranging, gives

$$m\omega^2 = k - b^2/4m.$$

This can only give a real value for ω if the right-hand side is positive, so

▶
$$4km > b^2$$

as required.
[This part of the problem could also be solved using the complex exponential notation. If we rewrite

$$x = A\exp(-\gamma t)\cos(\omega t + \phi)$$

as

$$x = \text{Re} \{ B \exp (i\alpha t) \},$$

where B and α can both be *complex*, the differentiations are much easier.

$$dx/dt = \text{Re} \{ Bi\alpha \exp (i\alpha t) \}$$

and

$$d^2x/dt^2 = \text{Re} \{ -B\alpha^2 \exp (i\alpha t) \}.$$

Substituting these results into the equation of motion gives

$$m\alpha^2 - i\alpha b - k = 0,$$

which is a quadratic in α. The solutions of this quadratic are

$$\alpha = \frac{ib \pm \sqrt{(-b^2 + 4mk)}}{2m}.$$

Now if α is a complex number $y + iz$, where y and z are real, the time-variation of x is given by

$$\text{Re} \{ \exp (iyt) \} \exp (-zt),$$

which is the required variation. The condition for α to have a non-zero real part is that the term under the square root sign should be positive, so

▶ $$4km > b^2$$

as before. Also, we can identify z as γ, so that

▶ $$\gamma = b/2m$$

as before.]

This variation of x with t corresponds to an under-damped decaying oscillation. The last part of the problem clearly deals with this kind of motion, so we know the form of differential equations we are looking for. Since the frequency f is 0.5 Hz, the angular frequency must be $2\pi f = \pi \, \text{s}^{-1}$, i.e.

$$\omega = \pi \quad \text{(in SI units)}.$$

Since the amplitude is halved in 2 s, we must have $\exp (-2\gamma) = 1/2$, so

$$\gamma = \tfrac{1}{2}\ln 2 \quad \text{(in SI units)}.$$

Now we know that $\gamma = b/2m$, so

$$b/m = \ln 2 \quad \text{(in SI units)}$$
$$= 0.6931.$$

We also know that

$$m\omega^2 = k - b^2/4m,$$

which can be rearranged to give

$$
\begin{aligned}
k/m &= \omega^2 + (b/2m)^2 \\
&= \pi^2 + (\ln 2)^2/4 \quad \text{(in SI units)} \\
&= 9.9897.
\end{aligned}
$$

Now the original differential equation for the motion of the object was

$$m\,d^2x/dt^2 + b\,dx/dt + kx = 0,$$

which can be divided by m to give

$$d^2x/dt^2 + (b/m)dx/dt + (k/m)x = 0,$$

so substituting our values for b/m and k/m we can finally write the differential equation as

▶ $$d^2x/dt^2 + 0.693dx/dt + 9.99x = 0 \quad \text{(in SI units)}.$$

[We have to specify that the coefficients are 'in SI units' since they are not dimensionless. The first coefficient has the dimensions of time^{-1} so it could be quoted as $0.693\ \text{s}^{-1}$, and the second coefficient has the dimensions of time^{-2} so it could be quoted as $9.99\ \text{s}^{-2}$.]

Problem 95

A steady force of 40 N is required to lift a mass of 2 kg vertically through water at a constant speed of $2\ \text{m s}^{-1}$. Assuming that the effect of viscosity can be described by a force proportional to velocity, determine the constant of proportionality. (The effect of buoyancy should be neglected.)

The same mass is then suspended in water by a spring with force constant $k = 100\ \text{N m}^{-1}$. Determine the equilibrium extension of the spring. The mass is released from rest 20 cm below its equilibrium position at time $t = 0$. Show that it will vibrate about the equilibrium position according to an equation of the form

$$d^2x/dt^2 + 2\gamma\,dx/dt + \omega_0^2 x = 0,$$

and determine γ and ω_0 for this system.

Given that $x = A\exp(-\gamma t)\cos(\omega t + \phi)$ is a solution of this equation if $\omega^2 = \omega_0^2 - \gamma^2$, calculate the period of oscillation of the system, and sketch the variation of x as a function of time.

Solution

When the mass is lifted vertically, it moves at constant speed so no net force must be acting upon it. The upward force of 40 N must therefore balance the sum of the downward forces, namely $mg + cv$ where m is the mass (2 kg), v is the velocity $(2\,\mathrm{m\,s^{-1}})$ and c is the constant of proportionality. Thus

▶
$$c = (40 - 2 \times 9.8)/2 = 10.2\,\mathrm{N\,s\,m^{-1}}.$$

When the mass is suspended in water by a spring of spring constant k, at equilibrium it will be stationary so no viscous force will act upon it. The weight mg of the mass will thus be entirely balanced by the tension kX in the spring, where X is its extension, so

▶
$$X = mg/k = 9.8 \times 2.0/100 = 0.196\,\mathrm{m}.$$

The general equation of motion of the mass about its equilibrium position is

$$m\,d^2x/dt^2 + c\,dx/dt + kx = 0,$$

which becomes, on dividing by m,

$$d^2x/dt^2 + (c/m)dx/dt + (k/m)x = 0.$$

▶
This has the required form if we identify $\gamma = c/2m$ and $\omega_0 = \sqrt{(k/m)}$. The numerical values of these coefficients are thus $\gamma = 2.55\,\mathrm{s^{-1}}$ and $\omega_0 = 7.07\,\mathrm{s^{-1}}$.

We are told that $x = A\exp(-\gamma t)\cos(\omega t + \phi)$ is a solution of this equation if $\omega^2 = \omega_0^2 - \gamma^2$, and ω is clearly the angular frequency of the oscillation. Substituting the values of γ and ω_0 gives

$$\omega = (7.07^2 - 2.55^2)^{1/2}\,\mathrm{s^{-1}} = 6.59\,\mathrm{s^{-1}},$$

▶
so the period of oscillation is $(2\pi/6.59)\,\mathrm{s} = 0.95\,\mathrm{s}$. This variation of x with t represents an exponentially decaying (damped) oscillation, which is plotted in figure 79 assuming $A = 1$ and $\phi = 0$. The amplitude of the oscillation falls by a factor of e in a time $1/\gamma = 0.39\,\mathrm{s}$, which is less than the period of oscillation, so the degree of damping is quite strong (Q is small—see problem 99), although it is still under-damped.

Problem 96

The equation of motion for a damped simple harmonic oscillator is

$$m\frac{d^2x}{dt^2} + k\frac{dx}{dt} + \lambda x = 0,$$

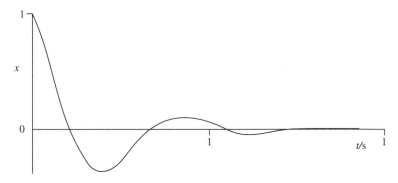

Figure 79

where k and λ are constants, m is the mass and x is the displacement of the system.

Describe the conditions for lightly damped, critically damped and over-damped oscillations. Draw diagrams to show how the displacement varies with time in the three cases, given that in each case the system is initially displaced and then released from rest.

A system whose natural frequency in the absence of damping is $4\,\text{rad}\,\text{s}^{-1}$ is subject to a damping force such that $k/m = 10\,\text{s}^{-1}$. Show that the system is over-damped and that the general solution for the displacement is

$$x = A\exp(-2t) + B\exp(-8t).$$

The mass is initially at $x = +0.5\,\text{m}$ and given an initial velocity V *towards* the equilibrium position. Find the smallest value of V that will produce a negative displacement.

Solution

As usual, we try a solution of the form

$$x = x_0 \exp(i\omega t)$$

and substitute into the differential equation of motion. After removing the constant factor $x_0 \exp(i\omega t)$ this gives the quadratic

$$m\omega^2 - ik\omega - \lambda = 0,$$

which is solved for ω to give

$$\omega = \frac{ik \pm \sqrt{(4m\lambda - k^2)}}{2m}.$$

The square root term determines the kind of damping involved:
  (a) Term under the square root is positive:

$$k^2 < 4m\lambda;$$
under-damped.

If $k^2 \ll 4m\lambda$, the system is lightly damped and the solution for ω is closely approximated by

$$\omega \approx \sqrt{\frac{\lambda}{m} + \frac{ik}{2m}},$$

where the real part describes oscillation at the natural frequency and the imaginary part describes exponential decay. We ignore the solution in which the real part of the frequency is negative, since it is not physically meaningful. However, there *are* two distinct solutions, and these can be combined by writing the displacement x as

$$x = (A\cos\omega_0 t + B\sin\omega_0 t)\exp(-\alpha t),$$

where A and B are constants, $\omega_0 = (\lambda/m - k^2/4m^2)^{1/2}$ and $\alpha = k/2m$.
  (b) Term under the square root is zero:

$$k^2 = 4m\lambda;$$
critically damped.

The solution is

$$\omega = \frac{ik}{2m},$$

i.e. exponential decay with no oscillation. The two solutions in this case are obtained by writing the general solution for x as

$$x = (A + Bt)\exp(-\alpha t),$$

where A and B are constants and $\alpha = k/2m$ as before.
  (c) Term under the square root is negative:

$$k^2 > 4m\lambda;$$
over-damped.

The solution of the quadratic equation now gives two imaginary values of ω, corresponding to exponential decays, without oscillation, with different time-constants. These two exponential decays can be combined to give a general solution for x:

$$x = A\exp(-\alpha t) + B\exp(-\beta t),$$

where A and B are constants,

$$\alpha = k/2m + (k^2/4m^2 - \lambda/m)^{1/2},$$

and

$$\beta = k/2m - (k^2/4m^2 - \lambda/m)^{1/2}.$$

These forms of behaviour are sketched in figure 80, for the cases $k^2/4m\lambda = 0$ (undamped), 1/4 (under-damped), 1 (critically damped) and 4 (over-damped).

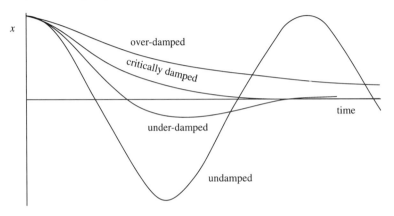

Figure 80

   In the numerical problem, we are given that the natural frequency is $4\,\text{s}^{-1}$ and that $k/m = 10\,\text{s}^{-1}$. Now since the natural frequency is given by $\sqrt{(\lambda/m)}$, we have

$$\lambda/m = 16$$

and

$$k/m = 10$$

in SI units. Thus the oscillation frequency ω is

$$\omega = \frac{ik \pm \sqrt{(4m\lambda - k^2)}}{2m}$$

$$= 5i \pm (16 - 25)^{1/2}$$
$$= 5i \pm 3i$$

and the two solutions are therefore $A\exp(-2t)$ and $B\exp(-8t)$, the system being over-damped. The general solution is thus

$$x = A\exp(-2t) + B\exp(-8t)$$

as required.

Now the initial conditions are that $x = +0.5$ and $dx/dt = -V$ (where V is a positive number) at $t = 0$, and these conditions will determine the values of A and B.

From the condition on x,

$$A + B = 0.5,$$

and since $dx/dt = -2A \exp(-2t) - 8B \exp(-8t)$, the condition on dx/dt gives

$$-2A - 8B = -V.$$

Combining these two relationships gives

$$A = (4 - V)/6,$$
$$B = (V - 1)/6,$$

so the subsequent motion of the mass is given by

$$x = \frac{4 - V}{6} \exp(-2t) + \frac{V - 1}{6} \exp(-8t).$$

For the displacement to become negative, it must pass through zero, so

$$\frac{4 - V}{6} \exp(-2t) = \frac{1 - V}{6} \exp(-8t),$$

thus

$$\frac{4 - V}{1 - V} = \exp(-6t).$$

If the displacement only just becomes negative, it will reach zero at $t = \infty$, so

$$\frac{4 - V}{1 - V} = 0.$$

Thus $V = 4$, i.e. the minimum value of V required to give a negative displacement is $4 \, \text{m s}^{-1}$.

Problem 97

A particle of mass m moves in one dimension and is subject to a restoring force which is proportional to its displacement x and a damping force which is proportional to its velocity. Derive the differential equation for its motion when it is also acted upon by a driving force $F_0 \cos \omega_F t$.

If $x = A \cos(\omega_F t + \phi)$, show that at low frequencies ω_F the phase ϕ is zero and the amplitude A is independent of the driving frequency ω_F, whereas at high frequencies $\phi = \pi$ and A depends on ω_F.

Solution

If the particle has position x and velocity dx/dt, the restoring force is $-kx$ (i.e. directed oppositely to the displacement) and the damping force is $-c\,dx/dt$ (opposite to the velocity), where k and c are constants. The net force acting on the particle is thus $F_0 \cos \omega_F t - kx - c\,dx/dt$, and this must be equal to the product of the particle's mass and its acceleration:

$$m\frac{d^2x}{dt^2} = F_0 \cos \omega_F t - kx - c\frac{dx}{dt},$$

which can be rearranged as

$$m\frac{d^2x}{dt^2} + c\frac{dx}{dt} + kx = F_0 \cos \omega_F t.$$

If we assume that the solution to this differential equation has the form given, namely $x = A \cos(\omega_F t + \phi)$, we can evaluate dx/dt and d^2x/dt^2 and substitute them into the equation to find A and ϕ:

$$\frac{dx}{dt} = -\omega_F A \sin(\omega_F t + \phi),$$

$$\frac{d^2x}{dt^2} = -\omega_F^2 A \cos(\omega_F t + \phi).$$

Substituting into the differential equation gives

$$-m\omega_F^2 A \cos(\omega_F t + \phi) - c\omega_F A \sin(\omega_F t + \phi) + kA \cos(\omega_F t + \phi)$$
$$= F_0 \cos \omega_F t.$$

At sufficiently low frequencies we can ignore the terms whose magnitudes are proportional to ω_F^2 and ω_F on the left-hand side, so that

$$kA \cos(\omega_F t + \phi) \approx F_0 \cos \omega_F t.$$

▶ The phase ϕ is thus equal to zero and the amplitude A is equal to F_0/k, independently of ω_F. At sufficiently high frequencies, the term in ω_F^2 dominates the left-hand side so that

$$-m\omega_F^2 A \cos(\omega_F t + \phi) \approx F_0 \cos \omega_F t.$$

Now since $-\cos(x) = \cos(x + \pi)$, we can arrange for the amplitude A to
▶ be positive by putting $\phi = \pi$. The amplitude A is then $F_0/m\omega_F^2$, so that it depends on ω_F, as required.

Problem 98

Show that the steady state complex amplitude of a damped oscillator driven by an external force $F \exp(i\omega t)$ is given by the expression

$$A = \frac{F}{M(\omega_0^2 - \omega^2) + i\omega b}.$$

Explain the meanings of the symbols used and the condition for resonance.

A machine of total mass 100 kg is supported by a spring resting on the floor and its motion is constrained to be in the vertical direction only. The system is lightly damped with a damping constant of $942 \, \mathrm{N \, s \, m^{-1}}$. The machine contains an eccentrically mounted shaft which, when rotating at an angular frequency ω, produces a vertical force on the system of $F_0\omega^2 \cos \omega t$, where F_0 is a constant. It is found that resonance occurs at 1200 r.p.m. (revolutions per minute) and the amplitude of vibration in the steady state is then 1 cm. Estimate the amplitude of vibration in the steady state when the driving frequency is (a) 2400 r.p.m. and (b) very large. You may assume that gravity has a negligible effect on the motion.

Solution

Assume that the system consists of a mass M connected to a spring of spring constant k and damping constant c. The equation of motion is thus

$$M d^2x/dt^2 + c \, dx/dt + kx = F \exp(i\omega t).$$

Assume that the solution for x is given by $x = A \exp(i\omega t)$, and substitute this into the differential equation. Eliminating $\exp(i\omega t)$, this gives

$$-MA\omega^2 + ic\omega A + kA = F,$$

and on rearranging, this gives

$$A = \frac{F}{k - M\omega^2 + i\omega c}.$$

▶ This has the required form if we identify b with c (the damping constant) and ω_0 with $\sqrt{(k/M)}$ (the resonant frequency if the damping term is zero).

Given that the force amplitude is $F_0\omega^2$, we may write the amplitude of vibration as

$$|A| = \frac{F_0\omega^2}{\sqrt{(M^2[\omega_0^2 - \omega^2]^2 + \omega^2 b^2)}}.$$

Since the system is lightly damped, its resonant frequency will be ω_0 at which value $|A|$ is

$$\frac{F_0 \omega_0}{b} = 1 \text{ cm}.$$

(a) At 2400 r.p.m., $\omega = 2\omega_0$, so

$$|A| = \frac{4F_0 \omega_0^2}{\sqrt{(9M^2 \omega_0^4 + 4\omega_0^2 b^2)}}$$

$$= \frac{4F_0 \omega_0}{b} \left(\frac{9M^2 \omega_0^4}{\omega_0^2 b^2} + 4 \right)^{-1/2}.$$

Taking $M = 100$ kg, $b = 942$ N s m^{-1}, $\omega_0 = 40\pi$ s^{-1} and $F_0 \omega_0 / b = 1$ cm, this gives

▶ $|A| = 0.10$ cm at 2400 r.p.m.

(b) As $\omega \to \infty$,

$$|A| \to \frac{F_0 \omega^2}{M \omega^2} = \frac{F_0}{M} = \frac{F_0 \omega_0}{b} \frac{b}{\omega_0 M},$$

▶   thus $|A|$ tends to a constant value of 0.075 cm.

Problem 99

Estimate the quality factor Q of the bell 'Big Ben'.

Solution

Q values relate to resonant systems, and are defined in terms of the differential equation governing the motion of the system. They can be approximated, however, by considering the behaviour of such systems at resonance or under conditions of freely decaying oscillation. The latter is clearly more appropriate here, so we will need to estimate the resonant frequency of the bell and the time for its amplitude to fall significantly, say by a factor of e.

We probably do not know (I certainly don't) the frequency at which the bell rings, but it is fairly low (below middle C, which is about 260 Hz) so we might guess 100 Hz (the lower limit of human audibility is about 20 Hz). The time required for the amplitude to decay by a factor of e we might estimate at 2 s.

A good approximation to Q for large values of Q is 2π times the number of cycles required for the amplitude to decay by a factor of e, so our estimate of Q is

$$(2\pi) \times (100) \times (2) \approx 1300.$$

And at this level of approximation, we would not be justified in quoting the answer any more precisely than

▶ $Q \approx 10^3.$

[Definitions of the quality factor Q vary, but the most precise is as a ratio of coefficients in the differential equation describing the motion of a resonant system. If the equation is

$$A\, d^2x/dt^2 + B\, dx/dt + Cx = F,$$

where F is the forcing function and x is the displacement,

$$Q = \frac{\sqrt{(AC)}}{B}.$$

Approximations to Q are:
   (i) ratio of amplitude at resonance to amplitude at zero frequency, for the same amplitude of force;
   (ii) ratio of resonant frequency to the separation of the two frequencies at which the amplitude is $1/\sqrt{2}$ times the maximum value, for the same amplitude of force;
   (iii) π divided by the logarithmic decrement Δ (defined as the natural logarithm of the ratio of the amplitudes of successive half-oscillations); this is equal to 2π times the number of oscillations required to reduce the amplitude by a factor of e;
   (iv) 2π divided by the fractional energy loss per cycle.
All of these approximations tend to Q when Q is large. None is particularly good for small Q, and (iv) is the worst of all.]

Problem 100

Explain why the displacement $A \sin(\mathbf{k.r} - \omega t)$ at position $\mathbf{r}$ and time t is called a plane wave. In which direction does it travel?
   A wave displacement is $y = 0.5 \sin(0.1x - 0.4t)$, where all quantities are in SI units. Determine
   (a) the wave amplitude,
   (b) the wavelength,
   (c) the time period,

(d) the wave velocity,
(e) the acceleration a of the displacement.
Sketch graphs of y and a at $t = 0$ for x in the range 0 to 100 m.

Solution

If the displacement is given by $A \sin(\mathbf{k.r} - \omega t)$, the phase of the wave is $\mathbf{k.r} - \omega t$. At a particular time t, a surface of constant phase is thus defined by $\mathbf{k.r} = $ constant, and this is the equation of a plane normal to
▶ the direction of the vector $\mathbf{k}$. Thus this displacement describes a simple harmonic wave propagating in the $\mathbf{k}$ direction.

$y = 0.5 \sin(0.1x - 0.4t)$ can be interpreted by comparing it with the standard form

$$y = A \sin(kx - \omega t)$$

for a wave propagating in the $+x$ direction.
▶ (a) by direct comparison, the amplitude A is 0.5 m.
(b) The wavenumber k is $0.1\ \mathrm{m}^{-1}$, so the wavelength
▶ $\lambda = 2\pi/k = 62.8$ m.
▶ (c) The angular frequency ω is $0.4\ \mathrm{s}^{-1}$, so the period
      $T = 2\pi/\omega = 15.7$ s.
▶ (d) The wave velocity v is $\omega/k = \lambda/T = 0.4/0.1 = 4\ \mathrm{m\,s}^{-1}$.
(e) The acceleration a is obtained by differentiating y twice with respect to time, thus:

$$a = \mathrm{d}^2 y/\mathrm{d}t^2 = -(0.4)^2(0.5)\sin(0.1x - 0.4t) \text{ in SI units}$$
▶ $$= -0.08\sin(0.1x - 0.4t).$$

In order to sketch y and a at $t = 0$, substitute this value of t into the expressions for y and a to obtain

$$y(t = 0) = 0.5\sin(0.1x),$$
$$a(t = 0) = -0.08\sin(0.1x).$$

These can now be plotted straightforwardly in figure 81.

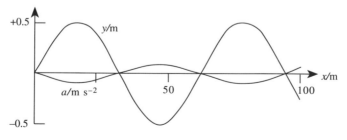

Figure 81

Problem 101

A wave is represented by

$$\psi_1 = 10\cos(5x + 25t),$$

where x is measured in metres and t in seconds. Show that this represents a travelling wave and deduce its wavelength, frequency, speed and direction of travel.

A second wave for which

$$\psi_2 = 20\cos(5x + 25t + \pi/3)$$

interferes with the first wave. Deduce the amplitude and phase of the resultant wave.

Solution

A harmonic wave with angular velocity ω travelling with velocity v in the $+x$ direction can be written as

$$\psi = A\cos\left(\frac{\omega x}{v} - \omega t - \phi\right) = A\cos\left(\omega t - \frac{\omega x}{v} + \phi\right),$$

where A is the amplitude of the wave and ϕ is its phase. Comparing this with the expression for ψ_1 shows that the wave is a harmonic travelling wave with $\omega = 25\,\text{s}^{-1}$ and $\omega/v = -5\,\text{m}^{-1}$. Thus $v = -5\,\text{ms}^{-1}$, so the wave is travelling at $5\,\text{ms}^{-1}$ in the negative x-direction. The angular frequency of the wave is $25\,\text{s}^{-1}$ so the frequency is $f = \omega/2\pi = 3.98\,\text{Hz}$. The wavelength λ is given by $v = f\lambda$ so $\lambda = 1.26\,\text{m}$.

We can find the resultant ψ of the two waves ψ_1 and ψ_2 from the principle of superposition:

$$\psi = \psi_1 + \psi_2.$$

To simplify the working, we will put $5x + 25t = K$ so that

$$\psi = 10\cos K + 20\cos(K + \pi/3).$$

We wish to express the resultant as $\psi = A\cos(K + \phi)$ where A is the amplitude and ϕ is the phase. Expanding both expressions and equating them gives

$$10\cos K + 20\cos K \cos\frac{\pi}{3} - 20\sin K \sin\frac{\pi}{3}$$
$$= A\cos K \cos\phi - A\sin K \sin\phi.$$

Since $\cos(\pi/3) = 1/2$ and $\sin(\pi/3) = (\sqrt{3})/2$, the left-hand side becomes $20 \cos K - 10\sqrt{3} \sin K$. Equating the coefficient of $\cos K$ and $\sin K$ gives

$$A \cos \phi = 20,$$
$$A \sin \phi = 10\sqrt{3}.$$

▶   Dividing the second result by the first gives $\tan \phi = (\sqrt{3})/2$ so $\phi = 0.714$ (radians) $= 40.9°$.

Squaring the two results and adding them gives $A^2 =$
▶ $20^2 + 10^2 \times 3 = 700$ so $A = 26.5$. Figure 82 shows these results graphically.

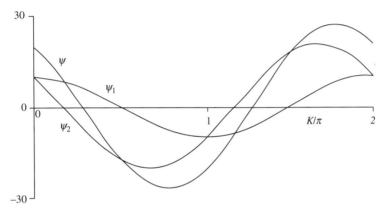

Figure 82

The figure confirms that the amplitude of ψ is 26.5. The phase ϕ can be determined from the figure by noting that ψ reaches its maximum value when $K = 1.773\pi$ which, since ψ has a period of 2π, is equivalent to $K = -0.227\pi$. Since $\cos(K + \phi)$ is maximum when $K = -\phi$, it follows that $\phi = 0.227\pi$ radians $= 0.714$ radians.

Problem 102

A short-wave (HF) radio receiver receives simultaneously two signals from a transmitter 500 km away, one by a path along the surface of the Earth, and one by reflexion from a portion of the ionospheric layer situated at a height of 200 km. The layer acts as a perfect horizontal reflector. When the frequency of the transmitted wave is 10 MHz it is observed that the combined signal strength varies from maximum to minimum and back to maximum 8 times per minute. With what slow vertical speed is the ionospheric layer moving? [Assume the Earth is flat and ignore atmospheric disturbances.]

Solution

Writing D for the length of the direct path along the Earth's surface and H for the height of the reflecting layer (see figure 83), we can see by Pythagoras' theorem that the path difference p between the two routes is

$$p = 2\left(H^2 + \frac{D^2}{4}\right)^{1/2} - D.$$

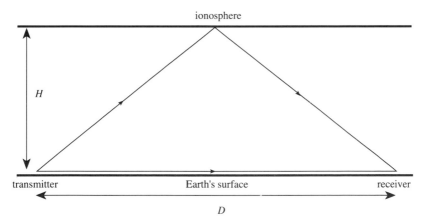

ionosphere

H

transmitter   Earth's surface   receiver

D

Figure 83

Interference between the signals arriving by the two routes causes the observed fluctuation in intensity, such that each time p changes by λ (the wavelength of the radiation) the received signal strength will vary through one cycle. The frequency f of the observed fluctuation will thus be given by

$$f = \frac{1}{\lambda}\frac{dp}{dt} = \frac{v}{c}\frac{dp}{dt},$$

where v is the frequency of the radiation and c is the speed of light. Differentiating our expression for p with respect to time gives

$$\frac{dp}{dt} = 2H\left(H^2 + \frac{D^2}{4}\right)^{-1/2}\frac{dH}{dt},$$

so we can write the fluctuation frequency as

$$f = \frac{2HvV}{c}\left(H^2 + \frac{D^2}{4}\right)^{-1/2},$$

where $V = dH/dt$ is the vertical velocity of the reflecting layer. Rearranging to make V the subject gives

$$V = \frac{fc}{2Hv}\left(H^2 + \frac{D^2}{4}\right)^{1/2}.$$

We can now substitute the values $H = 2 \times 10^5$ m, $D = 5 \times 10^5$ m, $v = 10^7$ Hz and $f = 8/60 = 0.133$ Hz to find $V = 3.2\,\mathrm{m\,s^{-1}}$. [The assumption that the Earth is flat makes a difference of less than 2% to the answer.]

Problem 103

Find the normal modes and frequencies of transverse waves on a string of length L which is under a tension T and is fixed at its two ends. If such a string with $L = 0.5$ m and mass per unit length $0.01\,\mathrm{kg\,m^{-1}}$ has a fundamental frequency of 247 Hz, what is the tension in the string?

Solution

The speed c of transverse waves on a string of mass per unit length ρ and tension T is

$$c = \sqrt{\frac{T}{\rho}}.$$

Now if the string is of length L and is fixed at its ends, each end must be stationary and hence a displacement node of the vibration.

The string must be an integral number of half-wavelengths in length, as shown in figure 84, so the wavenumber k_n of the nth normal mode must be given by

$$k_n = \frac{n\pi}{L}.$$

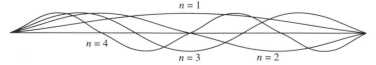

$n = 1$

$n = 4$

$n = 3$ $n = 2$

Figure 84

The angular frequency ω and the wavenumber k are related by

$$\omega = ck,$$

so the frequency of the nth mode is

$$f_n = \frac{ck_n}{2\pi},$$

thus

▶ $$f_n = \frac{n}{2L}\sqrt{\frac{T}{\rho}}$$

(Mersenne's law). Taking $n = 1$ (for the fundamental), and rearranging to make T the subject, we obtain

▶ $$T = 4L^2\rho f_1^2,$$

▶ so if $L = 0.5\,\text{m}$, $\rho = 0.01\,\text{kg}\,\text{m}^{-1}$ and $f_1 = 247\,\text{Hz}$, T must be 610 N. [These figures are typical of a piano string.]

Problem 104

A uniform inextensible string of length l and total mass M is suspended vertically and tapped at the top end so that a transverse impulse runs down it. At the same moment a body is released from rest and falls freely from the top of the string. How far from the bottom does the body pass the impulse?

Solution

Consider a point distance x above the bottom of the string. The mass of string below this point is clearly

$$Mx/l,$$

so the tension at this point is Mgx/l. The mass per unit length of the string is M/l, so the speed of transverse waves at the point x is

$$v = \sqrt{\frac{Mgx/l}{M/l}} = \sqrt{(gx)}.$$

The time $T(x)$ taken for the pulse to travel from $x' = l$ (the top) to x is thus

$$T(x) = \int_x^l \frac{dx'}{\sqrt{(gx')}}$$

$$= \left[2\sqrt{\frac{x'}{g}} \right]_x^l$$

$$= 2(l/g)^{1/2} - 2(x/g)^{1/2}.$$

The time $t(x)$ for a body to fall from rest through a distance $l - x$ is clearly

$$t(x) = \sqrt{\frac{2(l - x)}{g}}.$$

Equating $T(x)$ and $t(x)$ to find the value at which the pulse and the body meet, we obtain

$$\sqrt{\frac{2(l - x)}{g}} = 2\sqrt{\frac{l}{g}} - 2\sqrt{\frac{x}{g}}.$$

Eliminating $\sqrt{g}$, this gives

$$\sqrt{(2[l - x])} = 2\sqrt{l} - 2\sqrt{x}.$$

Squaring and dividing by 2:

$$l - x = 2l + 2x - 4\sqrt{(lx)}.$$

Rearranging:

$$4\sqrt{(lx)} = l + 3x.$$

Squaring again:

$$16lx = l^2 + 9x^2 + 6lx.$$

Rearrange again:

$$9x^2 - 10lx + l^2 = 0.$$

Solve the quadratic for x:

$$x = \frac{10l \pm \sqrt{(100l^2 - 36l^2)}}{18}$$

$$= l \text{ or } l/9.$$

▶ $x = l$ clearly represents the situation at the instant the pulse and the body are released, so they meet again a distance $l/9$ from the bottom of the string.

Problem 105

A long string of mass per unit length $0.2 \, \mathrm{kg \, m^{-1}}$ is stretched to a tension of 500 N. Find the speed of transverse waves on the string and the mean power required to maintain a travelling wave of amplitude 10 mm and wavelength 0.5 m.

The string is joined to another, of mass per unit length $0.8 \, \mathrm{kg \, m^{-1}}$. What fraction of the power carried by the wave is transmitted to the second string?

Solution

The speed c of transverse waves on a stretched string is given by

$$c = \sqrt{\frac{T}{\rho}},$$

where T is the tension and ρ is the mass per unit length. Setting $T = 500 \, \mathrm{N}$ and $\rho = 0.2 \, \mathrm{kg \, m^{-1}}$ gives

▶ $c = 50 \, \mathrm{m \, s^{-1}}.$

The power required to maintain a travelling wave of amplitude a and angular frequency ω is

$$P = \frac{c \rho a^2 \omega^2}{2},$$

so putting $c = (T/\rho)^{1/2}$ and $\omega = 2\pi c/\lambda$ this becomes

$$P = \frac{2\pi^2 a^2 T^{3/2}}{\lambda^2 \rho^{1/2}}.$$

▶   Substituting the values given, we obtain $P = 197 \, \mathrm{W}$.

[If we cannot remember the formula for the power required to maintain a wave on a stretched string it is fairly straightforward to derive. We can calculate the total energy per unit length by calculating the kinetic energy per unit length and doubling the result, since the average kinetic and potential energies in a travelling wave are equal:

Write

$$y = a \sin (kx - \omega t)$$

for the displacement of the string, and differentiate to find the velocity:

$$\partial y / \partial t = -a \omega \cos (kx - \omega t).$$

Thus at a particular instant, say $t = 0$, the velocity is given by $-a\omega\cos(kx)$.

Consider an element of length dx at position x along the string. Its mass is ρdx and its velocity is $-a\omega\cos(kx)$, so its kinetic energy is

$$\rho a^2 \omega^2 \cos^2(kx)dx/2.$$

Thus the total kinetic energy in length L is

$$\frac{1}{2}\rho a^2 \omega^2 \int_0^L \cos^2(kx)\,dx.$$

Now the average value of $\cos^2(kx)$ is $1/2$, so the kinetic energy per unit length must be

$$\tfrac{1}{4}\rho a^2 \omega^2.$$

The total energy (kinetic plus potential) per unit length is twice this value, and in order to maintain a travelling wave we must increase the length of string which is moving, at a rate of c (length per unit time). Thus the rate at which energy is being supplied to the wave, i.e. the power, is

$$2c\frac{1}{4}\rho a^2 \omega^2 = \frac{c\rho a^2 \omega^2}{2}.]$$

To calculate the power transmission coefficient into the second string, it is easiest to use wave impedances. The impedance Z of a transverse wave on a stretched string is given by

$$Z = c\rho = (T\rho)^{1/2},$$

so the impedance of the first string is $Z_1 = 10 \text{ kg s}^{-1}$. The mass per unit length of the second string is given as 0.8 kg m^{-1}, and the tension must be the same as that in the first string (since they are connected), so $Z_2 = 20 \text{ kg s}^{-1}$. The power transmission coefficient is given by

$$\frac{4Z_1 Z_2}{(Z_1 + Z_2)^2}$$
$$= 4 \times 10 \times 20/30^2$$
$$= 8/9.$$

[As before, if we cannot remember the formula relating the power transmission coefficient to the wave impedances of the strings, or the expression for the wave impedance of a stretched string, the result can be derived from first principles. We will assume that the strings are stretched along the x-axis, with the first (lighter) string at $x \leq 0$ and the second (heavier) string at $x \geqslant 0$. The incoming wave on the first string has amplitude a ($= 10$ mm) and is travelling in the $+x$ direction, so it can be described by the equation

$$y = a \exp(i[k_1 x - \omega t]),$$

where k_1 is the wavenumber of the waves on this string. This string also carries a reflected wave travelling in the $-x$ direction. If we assume that this wave has an amplitude ar, it can be expressed as

$$y = ar \exp(i[-k_1 x - \omega t]),$$

so that the total disturbance on the first string can be described by

$$y = a \exp(i[k_1 x - \omega t]) + ar \exp(i[-k_1 x - \omega t]) \quad \text{for } x \leqslant 0.$$

Let us suppose that the wave transmitted into the second string has amplitude at and wavenumber k_2. The wave is travelling in the $+x$ direction so it can be described by

$$y = at \exp(i[k_2 x - \omega t]) \quad \text{for } x \geqslant 0.$$

We assume that the second string is infinitely long so that we do not have to consider any reflected waves travelling in the $-x$ direction.

We can now consider the boundary conditions at $x = 0$ where the strings are joined. Firstly, we note that the y-coordinates of the two strings must be identical at all times at $x = 0$ (otherwise they would not be joined together), which gives

$$a + ar = at,$$

or

$$1 + r = t. \tag{1}$$

Secondly, we note that the tensions in the two strings are the same. This implies that the value of dy/dx must be the same for each string at $x = 0$ (otherwise there would be a kink in the string at the interface, resulting in a net force acting on the interface which has no mass). Evaluating dy/dx for string 1, we find

$$dy/dx = ik_1 a \exp(i[k_1 x - \omega t]) - ik_1 ar \exp(i[-k_1 x - \omega t]),$$

and for string 2,

$$dy/dx = ik_2 at \exp(i[k_2 x - \omega t]),$$

so if the values of dy/dx are to be equal at $x = 0$ we must have

$$ik_1 a - ik_1 ar = ik_2 at$$

or

$$1 - r = k_2 t / k_1. \tag{2}$$

Combining equations (1) and (2) to eliminate r and hence find t, we obtain

$$t = \frac{2k_1}{k_1 + k_2}.$$

Now the wavenumber k is related to the wave speed c by $\omega = ck$, where ω, the frequency of the wave, is constant. Thus $k_2/k_1 = c_1/c_2$, and since the wave speed c is proportional to $\rho^{-1/2}$ where ρ is the mass per unit length, $k_2/k_1 = (\rho_2/\rho_1)^{1/2}$. Taking $\rho_1 = 0.2 \text{ kg m}^{-1}$ and $\rho_2 = 0.8 \text{ kg m}^{-1}$ gives $k_2/k_1 = 2$, so $t = 2/3$. The amplitude of the wave transmitted into the second string is thus $2/3$ of the amplitude of the incident wave.

We have already shown that the power carried out by a wave is given by $c\rho a^2 \omega^2/2$, so if T and ω are constant, the power is proportional to $a^2 \rho^{1/2}$. The power transmitted into the second string is thus $(2/3)^2(4)^{1/2}$ times the incident power, so the power transmission coefficient is $8/9$, as before.]

Problem 106

The phase velocity of a surface wave on a liquid of surface tension T and density ρ is

$$v_{\mathrm{p}} = \sqrt{\left(\frac{g\lambda}{2\pi} + \frac{2\pi T}{\lambda \rho} \right)},$$

where g is the acceleration due to gravity and λ is the wavelength of the wave. Find the group velocity v_{g} of the surface waves.

What is v_{g} when v_{p} takes its minimum value as a function of wavelength?

Solution

The first thing we need to do is to rewrite the equation for v_{p} as a dispersion equation relating ω, the angular frequency, to k, the wavenumber. Recalling that

$$v_{\mathrm{p}} = \omega/k$$

and

$$\lambda = 2\pi/k,$$

we obtain

$$\frac{\omega}{k} = \sqrt{\left(\frac{g}{k} + \frac{kT}{\rho}\right)},$$

so

$$\omega = \sqrt{\left(gk + \frac{k^3 T}{\rho}\right)}.$$

Since $v_g = d\omega/dk$, we may differentiate this to find the group velocity:

$$v_g = \frac{d\omega}{dk} = \frac{g + \frac{3k^2 T}{\rho}}{2\sqrt{\left(gk + \frac{k^3 T}{\rho}\right)}}.$$

Since we have introduced the variable $k\ (= 2\pi/\lambda)$, not given in the problem, we should really rewrite this expression using λ instead of k:

▶ $$v_g = \frac{g + 12\pi^2 T/\rho\lambda^2}{2\sqrt{(2\pi g/\lambda + 8\pi^3 T/\rho\lambda^3)}}.$$

We are asked to evaluate v_g when v_p is minimum. We can save ourselves a little trouble by realising that when v_p is minimum, so is v_p^2, and the condition for this is that

$$\frac{d(v_p^2)}{d\lambda} = 0.$$

Now by differentiating the original expression for v_p, this gives

$$\frac{g}{2\pi} - \frac{2\pi T}{\lambda^2 \rho} = 0,$$

so

$$\lambda = 2\pi(T/g\rho)^{1/2}.$$

Substituting this value of λ back into the expression for v_g gives the value of v_g as

▶ $$v_g = \frac{g + 3g}{2\sqrt{\left(g\sqrt{\frac{\rho g}{T}} + g\sqrt{\frac{\rho g}{T}}\right)}} = \frac{2g}{\sqrt{(2\sqrt{[\rho g^3/T]})}} = \sqrt{2}\left(\frac{Tg}{\rho}\right)^{1/4}.$$

[The condition that $d(v_p^2)/d\lambda = 0$ really only defines an extremum (i.e. a maximum *or* minimum) of v_p^2, and hence of v_p. To convince ourselves that this is indeed a minimum, we can either differentiate again with respect to λ, and it is clear that this second derivative must be negative, or we can look back at the original expression for v_p. It is clear that at small values of λ, the term $2\pi T/\lambda\rho$ dominates, so that v_p becomes large at small λ. Similarly, at large values of λ the term $g\lambda/2\pi$ dominates, so that v_p is large at large values of λ. Thus the extremum which we have found must be a minimum, and not a maximum. This can be demonstrated by calculating v_p as a function of λ for water, for which we can take $\rho = 998\ \text{kg m}^{-3}$ and $T = 0.0727\ \text{N m}^{-1}$, as shown in figure 85.

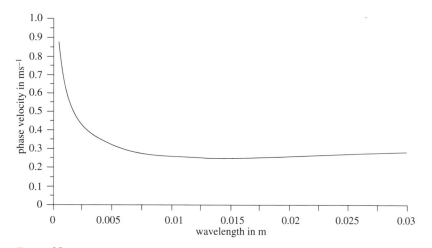

Figure 85

The minimum value of v_p occurs at a wavelength of 0.017 m and has a value of $0.23\ \text{m s}^{-1}$. The value of v_g at this wavelength is also equal to $0.23\ \text{m s}^{-1}$.]

Problem 107

The phase velocity v of gravity waves in a liquid of depth h is given by the formula

$$v^2 = \frac{g}{k}\tanh kh,$$

where g is the acceleration of free fall and $k = 2\pi/\lambda$ is the wavenumber, λ being the wavelength. Sketch the dispersion relation for such waves, and show that the group velocity is always between $v/2$ and v.

Find the phase and group velocities for gravity waves of frequency 1 Hz in a liquid of depth 0.1 m.

Solution

First we need to deduce the form of the dispersion relation $\omega(k)$, which we can do by using the fact that the phase velocity is ω/k. Substituting this into the expression we have been given for the phase velocity v, and rearranging, gives

$$\omega^2 = gk \tanh(kh).$$

In order to sketch this function (see figure 86), we can consider the limiting forms as $k \to 0$ and $k \to \infty$.

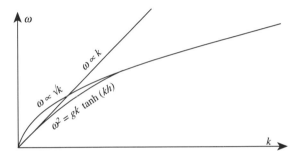

Figure 86

When $k \to 0$, $\tanh(kh) \to kh$ so $\omega \to (gh)^{1/2}k$, i.e. $\omega \propto k$.
When $k \to \infty$, $\tanh(kh) \to 1$ so $\omega \to (gk)^{1/2}$, i.e. $\omega \propto \sqrt{k}$.
The group velocity v_g is given by $d\omega/dk$, so if we differentiate our expression for ω^2 we obtain

$$2\omega v_g = g \tanh(kh) + \frac{gkh}{\cosh^2(kh)}.$$

Thus

$$\frac{v_g}{v} = \frac{k}{\omega} \frac{d\omega}{dk} = \frac{gk \tanh(kh) + \dfrac{gk^2 h}{\cosh^2(kh)}}{2gk \tanh(kh)}$$

$$= \frac{1}{2} + \frac{kh}{2 \sinh(kh) \cosh(kh)}$$

$$= \frac{1}{2} + \frac{kh}{\sinh(2kh)}.$$

The function $x/\sinh(2x)$ is shown in figure 87.

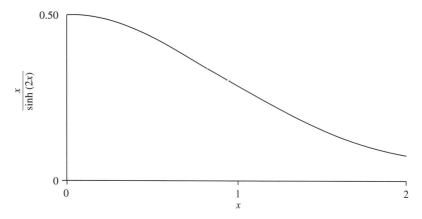

Figure 87

Now as $kh \to 0$, $\sinh(2kh) \to 2kh$ so $v_g/v \to 1$, and as $(kh) \to \infty$, $kh/\sinh(2kh) \to 0$ so $v_g/v \to 1/2$. Since $kh/\sinh(2kh)$ is a monotonic function, the value of v_g/v must always lie between these values, so the group velocity is always between $v/2$ and v as required.

At a frequency of 1 Hz the angular frequency is $2\pi\,\mathrm{s}^{-1}$. In order to find the phase and group velocities for waves of this frequency in a liquid of depth $h = 0.1$ m, we first need to find the corresponding value of k.

If we put $x = kh$, the dispersion relation can be written as

$$\frac{\omega^2 h}{gx} = \tanh x,$$

so the equation we have to solve is

$$\frac{0.402}{x} = \tanh x.$$

This can be solved graphically or numerically* to give $x = 0.680$, so $k = 6.80\,\mathrm{m}^{-1}$. The phase velocity is thus

$$\sqrt{\left(\frac{9.81}{6.80}\tanh 0.680\right)}$$

▶ $= 0.92\,\mathrm{m\,s}^{-1}$

and our expression for v_g/v gives

$$\frac{v_g}{v} = \frac{1}{2} + \frac{0.680}{\sinh 1.36} = 0.874,$$

▶ so the group velocity is $0.874 \times 0.92 = 0.80\,\mathrm{m\,s}^{-1}$.

[*To solve this equation numerically it is probably quickest to use the Newton–Raphson method. If we put

$$f(x) = x \tanh x - a = 0,$$

where $a = 0.402$, and x_n is an approximation to the root, then

$$x_{n+1} = x_n - \frac{f(x_n)}{f'(x_n)}$$

should be a better approximation. Since

$$f'(x) = \frac{x}{\cosh^2 x} + \tanh x,$$

this gives

$$x_{n+1} = x_n - \frac{x_n \tanh x_n - a}{\dfrac{x_n}{\cosh^2 x_n} + \tanh x_n} = \frac{x_n^2 + \dfrac{a}{2}(1 + \cosh 2x_n)}{x_n + \dfrac{\sinh 2x_n}{2}}.$$

A suitable starting value x_0 is $\sqrt{0.402} = 0.634$, since for small x, $\tanh x \approx x$. Thus

$$x_1 = 0.681,$$
$$x_2 = 0.680,$$
$$x_3 = 0.680,$$

so the method has converged to three significant figures after two steps.]

Problem 108

Prove that the group velocity v_g of electromagnetic waves in a dispersive medium with refractive index n is given by

$$v_g = \frac{c}{n + \omega \dfrac{dn}{d\omega}},$$

where c is the free-space velocity of light and ω is the angular frequency of the waves.

A *pulsar* is a star which emits very sharp pulses over a broad range of radio frequencies. These travel through the interstellar medium in a straight line before arriving at Earth. Radio observations show that the arrival time of a particular pulse measured at 400 MHz is 700 ms later than the arrival time of the same pulse measured at 1400 MHz. The refractive index of the interstellar medium is given by

$$n^2 = 1 - \frac{Ne^2}{\varepsilon_0 m \omega^2},$$

where e and m are the charge and mass of an electron, and N is the electron density, which is known to have an approximately uniform value of 3×10^4 m^{-3} in the space between the Earth and the pulsar. Calculate the distance to the pulsar.

Solution

The refractive index n is defined as the ratio of c to the phase velocity, so

$$n = \frac{ck}{\omega}.$$

The group velocity v_g is $d\omega/dk$, but since the required answer is given in terms of $dn/d\omega$ rather than dn/dk it will be more convenient to calculate the reciprocal of the group velocity:

$$\frac{1}{v_g} = \frac{dk}{d\omega} = \frac{1}{c}\frac{d}{d\omega}(n\omega)$$
$$= (n + \omega\, dn/d\omega)/c.$$

Thus

▶
$$v_g = \frac{c}{n + \omega\dfrac{dn}{d\omega}}$$

as required.

The group velocity is the velocity at which a pulse travels, so the time taken for a pulse to travel a distance D will be

$$t = \frac{D}{v_g} = \frac{D}{c}\left(n + \omega\frac{dn}{d\omega}\right).$$

This looks rather awkward given the expression for $n(\omega)$, so we should see whether the expression can be approximated. If we evaluate

$$\frac{Ne^2}{\varepsilon_0 m \omega^2}$$

at either of the two specified frequencies, say 400 MHz, we find that it is

$$\frac{3 \times 10^4 \times (1.6 \times 10^{-19})^2}{8.85 \times 10^{-12} \times 9.1 \times 10^{-31} \times (2\pi \times 4 \times 10^8)^2} = 1.5 \times 10^{-11}$$

i.e. very much less than unity. This means that a simple binomial expansion for n will be very accurate, so that

$$n \approx 1 - \frac{Ne^2}{2\varepsilon_0 m\omega^2}$$

and

$$\frac{dn}{d\omega} \approx \frac{2Ne^2}{2\varepsilon_0 m\omega^3}.$$

Thus

$$t = \frac{D}{v_g} \approx \frac{D}{c}\left(1 - \frac{Ne^2}{2\varepsilon_0 m\omega^2} + \frac{Ne^2}{\varepsilon_0 m\omega^2}\right) = \frac{D}{c}\left(1 + \frac{Ne^2}{2\varepsilon_0 m\omega^2}\right).$$

The component D/c is independent of ω and so will not contribute to the difference in arrival times of the two pulses, so we can write

$$\Delta t = \frac{Ne^2 D}{2\varepsilon_0 mc}\left(\frac{1}{\omega_1^2} - \frac{1}{\omega_2^2}\right)$$

for the difference in arrival times. This can be rearranged to give

$$D = \frac{2\varepsilon_0 mc\Delta t}{Ne^2}\left(\frac{1}{\omega_1^2} - \frac{1}{\omega_2^2}\right)^{-1}.$$

Taking $\omega_1 = 2\pi \times 4 \times 10^8 \text{ s}^{-1}$ and $\omega_2 = 2\pi \times 1.4 \times 10^9 \text{ s}^{-1}$, $\varepsilon_0 = 8.85 \times 10^{-12} \text{ F m}^{-1}$, $e = 1.6 \times 10^{-19} \text{ C}$, $N = 3 \times 10^4 \text{ m}^{-3}$, $c = 3 \times 10^8 \text{ m s}^{-1}$, $m = 9.1 \times 10^{-31} \text{ kg}$ and $\Delta t = 700 \text{ ms}$, this gives

▶ $$D = 3.0 \times 10^{19} \text{ m}.$$

Since a light-year is

$$3 \times 10^8 \times 60 \times 60 \times 24 \times 365 \text{ m}$$
$$= 9.5 \times 10^{15} \text{ m},$$

this distance is about 3000 light years. The diameter of the Galaxy is about 75 000 light years, so our answer is a reasonable one.

Optics

Problem 109

According to Descartes' theory of the formation of the rainbow, a ray of sunlight is refracted as it enters a spherical raindrop, undergoes a single internal reflexion, and is then refracted as it leaves the drop, as shown in figure 88.

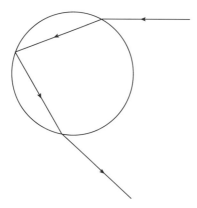

Figure 88

The rainbow is formed from the rays whose deviation from the original direction is either maximum or minimum. Show that the rainbow should form an arc of radius 42° round the point opposite to the Sun, and that it should have a width of approximately 1.6° with the red part on the outside. (The refractive index of water is 1.330 for red light and 1.341 for violet light.)

Solution

We will assume that the incident ray makes an angle θ with the normal to the drop's surface. The refracted ray will make an angle ϕ, where by Snell's law

$$\sin \phi = \frac{\sin \theta}{n}, \tag{1}$$

where n is the refractive index of water. Since the normal to any part of the surface of a sphere must pass through its centre, we can identify two congruent isosceles triangles, as shown in figure 89.

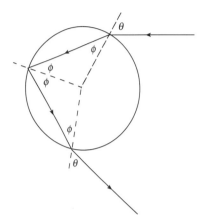

Figure 89

At the first refraction, the ray is deviated through an angle $\theta - \phi$ anticlockwise. At the internal reflexion it is deviated through $\pi - 2\phi$ in the same sense, and at the second refraction it is again deviated through $\theta - \phi$. The total deviation D is thus $2\theta - 4\phi + \pi$. To find the extremum, we set $dD/d\theta = 0$:

$$\frac{dD}{d\theta} = 2 - 4\frac{d\phi}{d\theta} = 0$$

$$\therefore \frac{d\phi}{d\theta} = \frac{1}{2}. \tag{2}$$

Differentiating equation (1) with respect to θ gives

$$\cos\phi\frac{d\phi}{d\theta} = \frac{\cos\theta}{n}$$

and substituting from (2) gives

$$\cos\phi = \frac{2\cos\theta}{n}. \tag{3}$$

We now have two equations, (1) and (3), relating ϕ and θ, so we can eliminate one of the angles to find an expression for the other one. The easiest way to do this is probably to add the squares of the two equations, using the fact that $\cos^2 \phi + \sin^2 \phi = 1$, to give

$$\frac{4\cos^2 \theta}{n^2} + \frac{\sin^2 \theta}{n^2} - 1.$$

To solve this, we can put $\cos^2 \theta = 1 - \sin^2 \theta$ and rearrange the resulting expression to give

$$\sin \theta = \sqrt{\frac{4 - n^2}{3}};$$

hence

$$\sin \phi = \sqrt{\frac{4 - n^2}{3n^2}}.$$

Taking $n = 1.330$ for red light gives $\theta = 59.585°$ and $\phi = 40.422°$, so the deviation is $2\theta - 4\phi + 180° = 137.48°$. Since a deviation of $180°$ corresponds to the sunlight's direction being reversed so that the light is observed to come from the direction opposite to the Sun, a deviation of $137.48°$ corresponds to a ring of red light of radius $180° - 137.48° = 42.52°$ centred on the antisolar point (see figure 90).

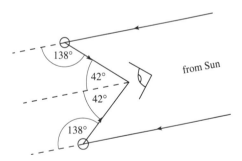

Figure 90

Repeating the calculation for violet light (for which $n = 1.341$) gives $\theta = 58.946°$ and $\phi = 39.705°$, from which it follows that the deviation is $139.07°$. Thus we have a ring of violet light of radius $40.93°$ centred on the antisolar point. The rainbow therefore has a mean angular radius of ▶ $(42.52 + 40.93)/2$ degrees $= 41.7°$ and an angular width of $(42.52 - 40.93)$ degrees $= 1.6°$, with the red part on the outside.

Problem 110

Show that the maximum magnification achievable with a magnifying glass of focal length f is $1 + D/f$, where D is the least distance of distinct vision.

Solution

Consider a converging lens forming a virtual image of an object of height h. Let the height of the image be h', the distance from the object to the lens u, and the distance from the lens to the virtual image x, as shown in figure 91.

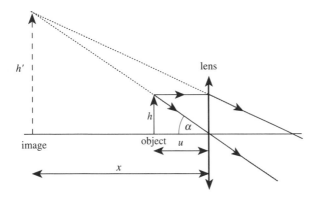

Figure 91

The angle subtended by the object at the lens is $\alpha = h/u$. (We are, as usual when considering rays and lenses, assuming that the rays always make small angles with the optic axis.) If the eye is placed next to the lens, this will also be the angle subtended by the image at the eye, but if the eye is placed at some distance from the lens the image will subtend a smaller angle. Thus the maximum magnification will be achieved if the eye is placed at the lens. The magnification is defined as the angle subtended by the image divided by the angle the object would subtend if it were observed, unaided, at the near point, so

$$M = \frac{\alpha}{h/D} = \frac{D}{u}.$$

For the maximum magnification, u must be made as small as possible. If it is made too small, however, x will be less than D and the eye will be unable to focus on the image. Thus we should express M in terms of x:

The relationship between u and x is given by

$$\frac{1}{u} + \frac{1}{v} = \frac{1}{f},$$

where $v = -x$ (because the image is virtual), so

$$\frac{1}{u} = \frac{1}{f} + \frac{1}{x}$$

and the magnification $M = D/u = D/f + D/x$. This clearly takes its maximum value when x takes its minimum value of D, giving

▶ $M_{max} = 1 + D/f$ as required.

Problem 111

The distance between an object and its real image, formed by a converging lens, is held fixed. Show that there are two possible positions for the lens, and that the size of the object is given by $(h_1 h_2)^{1/2}$ where h_1 and h_2 are the sizes of the two images.

Solution

Let us put f for the focal length of the lens, and u and v for the object and image distances respectively. u and v satisfy the equation

$$\frac{1}{u} + \frac{1}{v} = \frac{1}{f}$$

but we are also told that the distance from the object to the image is fixed. Let us call this distance a, so that

$$u + v = a.$$

We can combine these equations to eliminate either u or v: Multiplying the first equation through by uv to remove the fractions on the left-hand side gives

$$u + v = \frac{uv}{f}.$$

Substituting $u + v = a$ on the left-hand side and $v = a - u$ on the right-hand side gives

$$a = \frac{u(a - u)}{f},$$

which can be rearranged as a quadratic equation in u:

$$u^2 - au + af = 0.$$

This has two solutions

$$u = \frac{a \pm \sqrt{(a^2 - 4af)}}{2}$$

▶ provided that $a > 4f$ (which is the minimum distance between the object and the image). Thus we have shown that there are two possible positions of the lens for a given object-to-image distance a.

The magnification is given by $-v/u$, which we could write as $1 - a/u$ (using the fact that $u + v = a$). However, to evaluate this would involve taking the reciprocal of the expression we have just derived for u, which would be tedious. Instead, let us use the expression $1/u = 1/f - 1/v$ to write the magnification as $1 - v/f$, and then substitute for v to give the expression

$$-\frac{a - u - f}{f}$$

for the magnification. If the object height is h, the image heights are thus

$$h_1 = -\frac{h}{f}\left(a - \frac{a}{2} - \frac{\sqrt{[a^2 - 4af]}}{2} - f\right) = -\frac{h}{2f}(a - 2f - \sqrt{[a^2 - 4af]})$$

and

$$h_2 = -\frac{h}{f}\left(a - \frac{a}{2} + \frac{\sqrt{[a^2 - 4af]}}{2} - f\right) = -\frac{h}{2f}(a - 2f + \sqrt{[a^2 - 4af]}).$$

Forming the product $h_1 h_2$ gives

$$h_1 h_2 = \frac{h^2}{4f^2}(a^2 + 4f^2 - 4af - a^2 + 4af) = h^2,$$

▶ so we have shown that the object size is $(h_1 h_2)^{1/2}$ as required. [This technique is sometimes used to measure the size of an inaccessible object.]

Problem 112

Draw a scale diagram of the optical component in a refracting telescope in normal adjustment, giving the components the following values in your diagram:

objective lens	focal length 12 cm, diameter 6 cm;
single eyepiece lens	focal length 2 cm, diameter 2 cm.

Consider an object at infinity near the edge of the field of view but for which all light entering the objective will emerge from the eyepiece. Draw *three* rays (solid continuous lines) from this object; one through the centre and one through each edge of the objective, showing how they pass through the telescope and enter the observer's eye, situated at the exit-pupil of the telescope. Any construction rays should be dotted. Find the telescope's magnifying power, its field of view, and the diameter of the exit-pupil.

Solution

Since the object is at infinity, the objective must produce an image in its back focal plane, 12 cm behind it. The eyepiece is focussed on this image, so the distance from the objective to the eyepiece must be 14 cm. Knowing this, we can draw the lenses, optic axis and focal plane, as shown in figure 92.

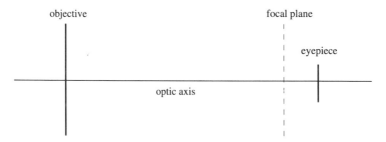

Figure 92

Now it is clear that the ray at the edge of the field of view which just passes through the eyepiece must connect (say) the lower edge of the objective to the upper edge of the eyepiece, so we can draw this ray ABC with B being the point at which the ray crosses the focal plane. B therefore defines the position of the internal image, and all rays from the object must pass through this point. We can thus draw a ray through B to the point O (the centre of the objective), and continue this to D (since rays passing through the centre of a lens are undeviated) and to L. We can also draw rays AF and HG parallel to OD, since these are the rays arriving at the objective from the object, and since the object is at infinity they must be parallel. Next, we can draw the ray HBK, and we find that the point K lies at the centre of the eyepiece lens, as shown in figure 93.

Since the ray HBK passes through the centre of the eyepiece lens it can be continued without deviation to J. Because the eyepiece is set up to produce an image at infinity (normal adjustment), rays LN and CM can be drawn parallel to KJ to give the complete diagram, shown in figure 94.

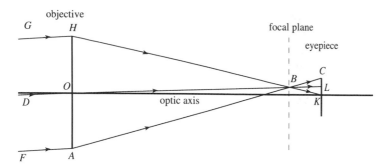

Figure 93

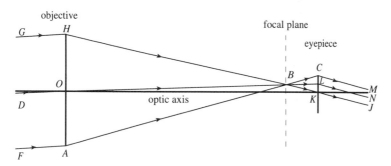

Figure 94

The magnification of the telescope is the ratio of the angle between the emerging rays and the optic axis to the angle between the incoming rays and the optic axis. Measuring (or calculating) the slope of the ray KJ shows it to be $-6/28$, and the slope of the ray DO is $+1/28$, so the magnification is -6.

If the angle between the incoming rays and the optic axis were any greater, not all of the rays entering the objective would pass through the eyepiece. The field of view of the telescope is thus $2 \arctan(1/28) = 4.1°$. It is clear that the diameter of the exit-pupil is the distance CK, and from our construction we have noted that K is on the optic axis of the system, so that the diameter is 1 cm. [We could also have calculated this by recalling that the exit-pupil diameter is given by the objective diameter divided by the modulus of the magnifying power.]

Problem 113

A beam of light of wavelength 600 nm passes through a slit of width 0.01 mm and strikes a screen, normal to the beam, placed a distance 2 m from the slit. Derive an expression for the shape of the intensity distribution seen on the screen and thereby deduce the width between the first minima of the distribution.

Suppose the screen were to be coated with phosphor and a beam of electrons were to be used instead of light. Through what potential difference should the electrons be accelerated before they reach the slit if the spreading of the beam is to be reduced by a factor of 10^3 below that found for light?

Solution

We may analyse this problem using the approximations of Fraunhofer diffraction.

Figure 95 shows the slit extending from $y = -w/2$ to $y = +w/2$, and two rays leaving the slit and arriving at the screen at position x. One of the rays leaves the slit at $y = 0$, and the other at y, and since $w \ll z$ (the distance from the slit to the screen), the two rays are very nearly parallel and the path difference between them is approximately $y \sin \theta$. The contribution to the wave amplitude at x due to the region of the slit extending from y to $y + dy$ thus has phase $ky \sin \theta$, where k is the wavenumber of the radiation, and amplitude proportional to dy (we are neglecting variations in the lengths of the rays), so we may write the amplitude in the direction θ as follows:

$$a(\theta) = \int_{-w/2}^{w/2} \exp(iky \sin \theta)\, dy.$$

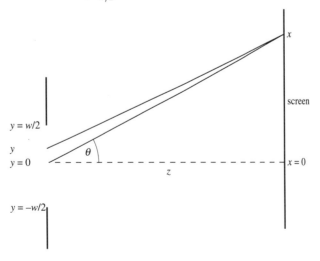

Figure 95

Evaluating the integral (which is a Fourier transform) gives

$$a(\theta) = \left[\frac{\exp{(iky\sin\theta)}}{ik\sin\theta}\right]_{-w/2}^{w/2} = \frac{2\sin\left(\dfrac{kw\sin\theta}{2}\right)}{k\sin\theta}.$$

Squaring this to find the intensity $I(\theta)$, and omitting constant factors, we obtain

$$I(\theta) \propto \frac{\sin^2\left(\dfrac{kw\sin\theta}{2}\right)}{\sin^2\theta}.$$

This function is shown in figure 96.

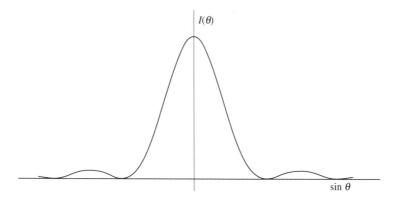

Figure 96

The first zeroes occur when

$$\frac{kw\sin\theta}{2} = \pm\pi \quad \therefore \sin\theta = \pm\frac{2\pi}{kw} = \pm\frac{\lambda}{w}.$$

Taking $\lambda = 600$ nm and $w = 0.01$ mm gives $\sin\theta = \pm0.06$, so $\theta = \pm0.060\,04$ radians. The distance x on the screen is given by $x = z\tan\theta$, so putting $z = 2$ m gives $x = \pm0.1202$ m. The width between the first zeroes is thus 240 mm.

If we require the spread to be reduced by a factor of 1000 using the same slit, it is clear from our formula that the wavelength must be reduced by a factor of 1000. Thus the electron beam should have a wavelength of 0.6 nm. Now we know that the wavelength of a particle is given by the de Broglie relation as

$$\lambda = \frac{h}{p},$$

where h is Planck's constant and p is the particle's momentum, and the momentum is related to the kinetic energy E for a non-relativistic particle of mass m by

$$E = \frac{p^2}{2m}.$$

If we put $E = eV$ where V is the accelerating voltage, and combine these expressions, we obtain

$$V = \frac{h^2}{2me\lambda^2}.$$

▶ Substituting $h = 6.6 \times 10^{-34}$ J s, $m = 9.1 \times 10^{-31}$ kg, $e = 1.6 \times 10^{-19}$ C and $\lambda = 6 \times 10^{-10}$ m gives $V = 4.2$ V.

Problem 114

A Young's slits experiment is set up in which two narrow slits, of separation d, are illuminated by light of wavelength λ. The diffraction pattern is viewed on a screen at a distance D. Derive expressions for the intensity distribution on the screen and the separation of the fringes.

If $D = 1$ m, $\lambda = 600$ nm, and the distance from the centre of the fringe pattern to the 10th bright band on one side is 30 mm, calculate the separation d of the slits.

A film of transparent material is placed over one of the slits, and the displacement of the centre of the fringe pattern is observed to be 30 mm. Calculate the refractive index of the material if its thickness is 20 μm.

Solution

We can use the same approach as in the previous problem. This time, however, instead of integrating the contributions to $a(\theta)$ over a range of y, we just add contributions from $y = +d/2$ and $y = -d/2$ to represent the light emerging from the two slits. Thus we have

$$a(\theta) = \exp\left(-\frac{ikd\sin\theta}{2}\right) + \exp\left(\frac{ikd\sin\theta}{2}\right) = 2\cos\left(\frac{kd\sin\theta}{2}\right).$$

Squaring this to find the intensity, and ignoring constant factors, gives

$$I(\theta) \propto \cos^2\left(\frac{kd\sin\theta}{2}\right).$$

Now if $\theta \ll 1$, we may take $\sin\theta \approx \tan\theta = x/D$, where x is the position on the screen. Substituting this expression for $\sin\theta$, and putting $\lambda = 2\pi/k$, gives the following expression for the intensity as a function of x:

▶
$$I(x) \propto \cos^2\left(\frac{\pi d x}{\lambda D}\right).$$

These 'cos² fringes' are sketched in figure 97.

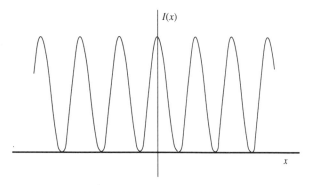

Figure 97

Since $\cos^2(n\pi) = 1$ for any integer n, the separation of the fringes is given by the value of x for which

$$\frac{\pi d x}{\lambda D} = \pi,$$

▶ i.e. the separation of the fringes is $\lambda D/d$. The position of the 10th bright fringe is thus $x = 10\lambda D/d$, so taking $\lambda = 600$ nm, $D = 1$ m and
▶ $10\lambda D/d = 30$ mm gives $d = 0.20$ mm.

When a sheet of transparent material of thickness t and refractive index n is placed in front of one of the slits, it changes the phase of the light passing through it by an amount

$$\Delta\phi = \frac{2\pi t}{\lambda}(n-1)$$

(provided the angle θ is still small). We can write the amplitude $a(\theta)$ in this case as

$$a(\theta) = \exp(-i\psi) + \exp(i\psi + i\Delta\phi),$$

where $\psi = (kd\sin\theta)/2$. The real part of this expression is

$$\cos\psi + \cos(\psi + \Delta\phi)$$

If we increase the wavelength slightly, the scale of the diffraction pattern, which is proportional to λ, will increase. We will assume that two wavelengths are resolved if the maximum of one coincides with the first zero of the other, as shown in figure 100.

diffraction pattern amplitude

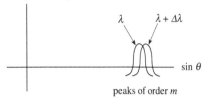

Figure 100

It is clear that the two peaks are separated by a distance λ/Nd in $\sin\theta$, and since the value of $\sin\theta$ is $m\lambda/d$, the fractional change in $\sin\theta$ is $1/mN$. Since $\sin\theta$ is proportional to λ, this must also be the fractional change in λ, i.e. $\Delta\lambda/\lambda$. The resolving power R is thus equal to mN, as required.

Problem 116

A telescope is used to observe at a distance of 10 km two objects which are 0.12 m apart and illuminated by light of wavelength 600 nm. Estimate the diameter of the objective lens of the telescope if it can just resolve the two objects.

Solution

Rayleigh's criterion for the angular resolution $\Delta\theta$ of a circular aperture of diameter $D \gg \lambda$ gives

$$\Delta\theta = 1.22\frac{\lambda}{D}$$

and for two objects at distance z separated by s the angular separation is

$$\Delta\theta = \frac{s}{z} \quad \text{(provided } s \ll z\text{)}.$$

Combining these two expressions gives

$$D = 1.22\frac{\lambda z}{s}$$

and substitution of $\lambda = 6 \times 10^{-7}$ m, $z = 10^4$ m and $s = 0.12$ m gives
▶ $D = 6$ cm.

Problem 117

The visibility V of the fringes observed in a Michelson interferometer is
defined as

$$V = \frac{I_{max} - I_{min}}{I_{max} + I_{min}},$$

where I_{max} and I_{min} are the intensities at the maxima and minima of the
fringe pattern. In observations of the sodium D-lines, V varies from a
maximum of 1.00 to a minimum of 0.33 over a range of approximately
500 fringes. Explain these observations and deduce values for the
fractional wavelength difference between the sodium D-lines and for their
relative intensities.

Solution

Let us consider first a Michelson interferometer being used to study light
of amplitude a and a single wavenumber k. The signal is split into two
equal parts of amplitude $a/2$, and one half is subject to a path difference d
with respect to the other half. When the two signals are recombined, the
complex amplitude of the resultant signal can be written as

$$\frac{a}{2} + \frac{a}{2} \exp[ikd] = \frac{a}{2}(1 + \exp[ikd]).$$

The intensity of the resultant is found by squaring and adding the real and
imaginary parts of this expression:

$$I = \frac{a^2}{4}[(1 + \cos[kd])^2 + \sin^2[kd]]$$

$$= \frac{a^2}{4}(2 + 2\cos[kd]) = \frac{I_0}{2}(1 + \cos[kd]),$$

where we have written I_0 for the intensity a^2 of the original signal. This
represents fringes whose intensity varies between zero and I_0, so the
visibility V of the fringes is one. The spacing of the fringes (which we can
define as the change in d between one maximum of intensity and the
next) is $2\pi/k$.

Now we can consider the output from the interferometer if the incident
light contains two closely spaced spectral components, the first having
intensity I_1 and wavenumber $k_0 - \Delta k/2$ and the second having intensity

I_2 and wavenumber $k_0 + \Delta k/2$. The intensity I of the output when the path difference is d is

$$I = \frac{I_1}{2}\left(1 + \cos\left[k_0 - \frac{\Delta k}{2}d\right]\right) + \frac{I_2}{2}\left(1 + \cos\left[k_0 + \frac{\Delta k}{2}d\right]\right),$$

which can be rewritten as

$$I = \frac{I_1 + I_2}{2} + \frac{I_1 + I_2}{2}\cos\left(\frac{\Delta k}{2}d\right)\cos(k_0 d)$$
$$+ \frac{I_1 - I_2}{2}\sin\left(\frac{\Delta k}{2}d\right)\sin(k_0 d).$$

This expression has the form

$$I = A + B\cos(k_0 d) + C\sin(k_0 d),$$

where A is a constant and B and C are slowly varying functions of d (since we are assuming that $\Delta k \ll k_0$). This describes intensity fringes with wavenumber k_0, varying from a minimum value of $A - \sqrt{(B^2 + C^2)}$ to a maximum value of $A + \sqrt{(B^2 + C^2)}$. This is shown schematically in figure 101.

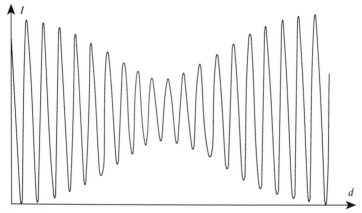

Figure 101

The visibility of the fringes is thus

$$V = \frac{\sqrt{(B^2 + C^2)}}{A}$$
$$= \frac{\sqrt{\left(\left[\frac{I_1 + I_2}{2}\right]^2\cos^2\left[\frac{\Delta k}{2}d\right] + \left[\frac{I_1 - I_2}{2}\right]^2\sin^2\left[\frac{\Delta k}{2}d\right]\right)}}{\frac{I_1 + I_2}{2}}.$$

The maximum value of V occurs when $d = 0$ and is clearly given by $V = 1$ as observed. The minimum value occurs when $\Delta k\, d/2 = \pi/2$ and is given by

$$V_{min} = \frac{I_1 - I_2}{I_1 + I_2}.$$

(In fact, it is given by the modulus of this expression. We are assuming that $I_1 > I_2$.) Since the observed minimum value of V is 0.33, the ratio of the intensities of the two lines must be $I_2/I_1 = 0.50$. We can calculate the fraction wavelength separation of the two lines as follows:

The first minimum in the visibility function occurs at $d = \pi/\Delta k$, but the fringe separation is $2\pi/k_0$ so the number of fringes N over which the visibility changes from maximum to minimum is given by

$$N = \frac{\pi}{\Delta k} \div \frac{2\pi}{k_0} = \frac{k_0}{2\Delta k}.$$

Since $k \propto 1/\lambda$, the fractional wavelength difference $\Delta\lambda/\lambda$ is equal to the fractional wavenumber difference $\Delta k/k$ (apart from a minus sign which we can clearly ignore since we are not interested in the sign of the difference), so

$$\frac{\Delta\lambda}{\lambda} = \frac{1}{2N}.$$

Taking $N = 500$ gives $\Delta\lambda/\lambda = 1/1000$. [In fact, the mean wavelength of the sodium D-lines is 589.30 nm and their separation is 0.59 nm, so the result is a very accurate one.]

Electromagnetism

Problem 118

The electric potential at a perpendicular distance r from a long straight wire of cross-sectional radius a is given by

$$V(r) = -K \ln \frac{r}{a},$$

where K is a constant. Calculate the electric field as a function of distance. Hence, using Gauss's theorem, determine the charge q per unit length of the wire.

A second identical wire, carrying charge $-q$ per unit length, is placed parallel to the first at a distance d from it. Calculate the potential difference between the wires, assuming that $d \gg a$.

Solution

In general, the relationship between the electric field E and the potential V is

$$\mathbf{E} = -\nabla V.$$

However, the field due to the wire clearly has cylindrical symmetry so we can write this as $E = -dV/dr$. Differentiating the expression for V gives
▶ $E = K/r$.

Gauss's theorem states that, for a closed region of space,

$$\int \mathbf{E} \cdot d\mathbf{s} = \frac{1}{\varepsilon_0} \sum Q,$$

where $\mathbf{E}$ is the electric field vector on the surface of the region, $d\mathbf{s}$ is an element of the vector area of the surface (pointing outwards), and $\sum Q$ is the total charge contained within the region of space. If we choose a cylindrical region of space, coaxial with the wire and having radius r and length L (as shown in figure 102), it is clear that over the curved surface $\mathbf{E}$ is constant and everywhere perpendicular to the surface.

The contribution to the integral is thus $2\pi r L E$. Over the ends of the cylinder, the electric field is normal to the surface so there is no

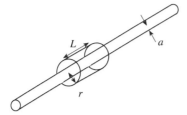

Figure 102

contribution to the integral. Thus, from Gauss's theorem, the charge contained within the cylinder is

$$Q = 2\pi r L E \varepsilon_0 = 2\pi K L \varepsilon_0,$$

▶ so the charge per unit length of the wire is $q = 2\pi K \varepsilon_0$.

If we now include another identical wire, carrying charge per unit length $-q$, distance d from the first (as shown in figure 103), we can find the total field by superposition.

Figure 103

The field at r due to the wire carrying charge per unit length $+q$ is K/r, and the field due to the wire carrying charge per unit length $-q$ is $K/(d-r)$. Thus the total field is

$$E = K\left(\frac{1}{r} + \frac{1}{d-r}\right).$$

The potential difference between the wires is found by integrating the field with respect to r, from $r = a$ to $r = d - a$:

$$V = K\int_a^{d-a}\left(\frac{1}{r} + \frac{1}{d-r}\right) dr = K[\ln r - \ln(d-r)]_a^{d-a}$$

$$= 2K \ln\frac{d-a}{a}.$$

[This result could also have been written down directly using the form of ▶ $V(r)$.] If $d \gg a$, V can be approximated as $V \approx 2K \ln(d/a)$.

Problem 119

A soap bubble 10 cm in radius with a wall thickness of 3.3×10^{-6} cm is charged to a potential of 100 V. The bubble bursts and falls as a spherical drop. Estimate the potential of the drop.

Solution

In order to find the charge on the bubble, we need to know the capacitance of a spherical shell of radius a.

Let us assume that such a shell is given a charge Q, and construct a Gaussian surface round it. Since the field is (by symmetry) radial, a spherical Gaussian surface of radius r $(> a)$ will be everywhere normal to the field. Gauss's theorem then gives $4\pi r^2 E = Q/\varepsilon_0$, so

$$E = \frac{Q}{4\pi r^2 \varepsilon_0}.$$

Since for a radial field $E = -dV/dr$, we can find the potential of the shell by integrating $-E$ with respect to r from $r = \infty$ to $r = a$:

$$V = -\frac{Q}{4\pi\varepsilon_0}\int_\infty^a \frac{dr}{r^2} = \frac{Q}{4\pi\varepsilon_0 a}.$$

Using the definition $Q = CV$ to find the capacitance C gives $C = 4\pi\varepsilon_0 a$.

If the initial radius of the bubble is a and it is charged to a potential V, it must therefore carry a charge $4\pi\varepsilon_0 aV$.

When the bubble bursts, it forms a spherical drop of radius a' carrying the same charge. The volume of this drop must be equal to the volume of liquid contained in the bubble, which is clearly $4\pi a^2 t$, where t is the wall thickness. Thus

$$4\pi a^2 t = \tfrac{4}{3}\pi a'^3 \text{ so } a' = (3a^2 t)^{1/3}.$$

Now if we assume that the soap solution conducts electricity, all of the charge will reside on the surface of the drop and its capacitance will be given by $4\pi\varepsilon_0 a'$ (see figure 104).

The potential of the drop will thus be

$$\frac{Q}{4\pi\varepsilon_0 a'} = \frac{4\pi\varepsilon_0 aV}{4\pi\varepsilon_0 a'} = \frac{aV}{a'} = \frac{a}{(3a^2 t)^{1/3}}V.$$

Putting $a = 0.10$ m, $t = 3.3 \times 10^{-8}$ m and $V = 100$ V gives the potential on the drop as 10 kV.

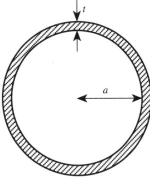

Bubble: $C = 4\pi\varepsilon_0 a$ Drop: $C = 4\pi\varepsilon_0 a'$

Figure 104

Problem 120

A spherical nucleus has a total charge Q (uniformly distributed) and radius R. Find the electric field at any point inside the nucleus at a distance r from the centre. Hence find the potential difference between the centre of the nucleus and its surface.

Solution

The field at radius r will, by Gauss's theorem, be that due to the charge enclosed within that radius. Since the charge is uniformly distributed, the enclosed charge will be proportional to r^3, and so equal to

$$Q\left(\frac{r}{R}\right)^3.$$

Thus the field at radius r ($<R$) is given by

$$E = Q\left(\frac{r}{R}\right)^3 \frac{1}{4\pi\varepsilon_0 r^2} = \frac{Qr}{4\pi\varepsilon_0 R^3}.$$

Since the field is radial, we can find the potential difference between the centre and the surface by integrating E with respect to r from $r = 0$ to $r = R$:

$$V = \frac{Q}{4\pi\varepsilon_0 R^3}\int_0^R r\, dr = \frac{Q}{4\pi\varepsilon_0 R^3}\frac{R^2}{2} = \frac{Q}{8\pi\varepsilon_0 R}.$$

[It is interesting to estimate an order of magnitude for this potential difference by putting $Q = 1.6 \times 10^{-19}$ C and $R = 10^{-14}$ m. This gives V of the order of 0.1 MV.]

Problem 121

Determine the electric field as a function of radius r in a spherically symmetric model of an atom in which the nucleus is a point charge of magnitude $+e$ and the electron charge is distributed with charge per unit volume $\rho(r)$ given by

$$\rho(r) = \left\{ \begin{array}{l} -\dfrac{15e}{8\pi a^3}\left(1 - \dfrac{r^2}{a^2}\right) \text{ for } r \le a, \\[2mm] 0 \text{ for } r > a, \end{array} \right.$$

where a is a constant.

Solution

For a spherically symmetric distribution of charge, the field $E(r)$ at radius r is given by Gauss's theorem as

$$E(r) = \frac{q(<r)}{4\pi\varepsilon_0 r^2},$$

where $q(<r)$ is the charge contained within a sphere of radius r. $q(<r)$ can be found by integrating the electron charge density with respect to volume, and including the nuclear charge:

$$q(<r) = e + 4\pi\int_0^r \rho r^2 \, dr$$

$$= e - \frac{15e}{2a^3}\int_0^r \left(1 - \frac{r^2}{a^2}\right)r^2 \, dr.$$

Performing the integral, we obtain

$$q(<r) = e - \frac{15e}{2a^3}\left(\frac{r^3}{3} - \frac{r^5}{5a^2}\right) = e\left(1 - \frac{5r^3}{2a^3} + \frac{3r^5}{2a^5}\right),$$

so that $E(r)$ is

$$E(r) = \frac{e}{4\pi\varepsilon_0 r^2}\left(1 - \frac{5r^3}{2a^3} + \frac{3r^5}{2a^5}\right).$$

This expression is only valid in the range $0 < r \le a$, since this is the range of values of r for which the expression for $\rho(r)$ is valid. Our expression for $q(<r)$ shows that it is zero when $r = a$, so $q(<r)$ must be zero for all values of $r \ge a$ (as we would expect). Thus $E(r)$ is also zero for $r \ge a$ (i.e. outside the atom).

Problem 122

Show that the potential energy of a charge Q uniformly distributed throughout a sphere of radius R is

$$\frac{3}{5} \frac{Q^2}{4\pi\varepsilon_0 R}.$$

Solution

The simplest way to solve this is to build up the sphere by adding a shell of radius dr to a sphere of radius r, and integrate the work done as the sphere is built up from $r = 0$ to $r = R$.

Since the charge is uniformly distributed, the charge on a sphere of radius r is

$$Q_1 = \frac{Qr^3}{R^3}.$$

The charge contained in a spherical shell of radius r and thickness dr is

$$Q_2 = Q\frac{4\pi r^2 dr}{\frac{4}{3}\pi R^3} = \frac{3Qr^2 dr}{R^3}.$$

The work done in bringing charges Q_1 and Q_2 from infinite separation to separation r is

$$\frac{Q_1 Q_2}{4\pi\varepsilon_0 r},$$

so we can write the increase dE in the potential energy resulting from adding a thickness dr to the sphere as

$$dE = \frac{Qr^3}{R^3}\frac{3Qr^2 dr}{R^3}\frac{1}{4\pi\varepsilon_0 r} = \frac{3Q^2 r^4}{4\pi\varepsilon_0 R^6}dr.$$

Integrating this from $r = 0$ to $r = R$ gives

$$E = \frac{3Q^2}{4\pi\varepsilon_0 R^6}\int_0^R r^4\, dr = \frac{3}{5}\frac{Q^2}{4\pi\varepsilon_0 R}$$

as required.

Problem 123

Show that the maximum values of the electric field $|E|$ for points on the axis of a uniform ring of charge q and radius a occur at a distance $x = \pm a/\sqrt{2}$.

If an electron is placed at the centre of the ring and is then displaced a small distance x along the axis ($x \ll a$), show that it oscillates with a frequency

$$v = \sqrt{\frac{eq}{16\pi^3 \varepsilon_0 a^3 m}},$$

where m is the mass of the electron.

Solution

To find the electric field at a point on the axis, we can use the expression for the field due to a point charge, and integrate round the ring as though it were a series of point charges.

Consider an element of the ring of length dl, as shown in figure 105. It contains a charge

$$\frac{dl}{2\pi a} q,$$

so at distance r it causes an electric field

$$dE = \frac{dl}{2\pi a} q \frac{1}{4\pi\varepsilon_0 r^2} = \frac{q \, dl}{8\pi^2 a \varepsilon_0 r^2}.$$

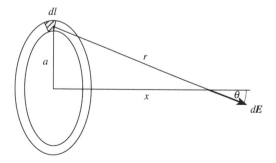

Figure 105

The field dE acts in the direction θ, and it can be resolved into a component $dE \cos \theta$ along the axis and a component $dE \sin \theta$ perpendicular to it. When we integrate round the whole ring, the perpendicular components will cancel out by symmetry, so we need only consider the components parallel to the axis. Thus

$$E = \frac{q}{8\pi^2 a \varepsilon_0 r^2} \cos \theta \int_0^{2\pi a} dl = \frac{q \cos \theta}{4\pi \varepsilon_0 r^2}.$$

Now we can express both $\cos \theta$ and r^2 in terms of x and a:

$$r^2 = a^2 + x^2,$$

$$\cos \theta = \frac{x}{r} = \frac{x}{(a^2 + x^2)^{1/2}},$$

so we can write E as a function of x as follows:

$$E = \frac{qx}{4\pi \varepsilon_0 (a^2 + x^2)^{3/2}}.$$

The form of $E(x)$ is sketched in figure 106.

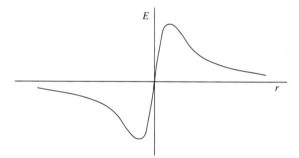

Figure 106

In order to find the points where $|E|$ reaches its maximum value we differentiate it with respect to x and find where $dE/dx = 0$:

$$\frac{dE}{dx} = \frac{q}{4\pi \varepsilon_0}\left([a^2 + x^2]^{-3/2} - 3x^2[a^2 + x^2]^{-5/2}\right),$$

so the condition that $dE/dx = 0$ requires that

$$(a^2 + x^2)^{-3/2} = 3x^2(a^2 + x^2)^{-5/2},$$
i.e. $a^2 + x^2 = 3x^2$,

▶ which gives $x = \pm a/\sqrt{2}$ as required.

Considering the motion of an electron near the centre of the ring, we can see that the field E varies approximately linearly with x:

$$E \approx \frac{qx}{4\pi\varepsilon_0 a^3}.$$

Since the electron has charge $-e$, we can write the force acting upon it as

$$F = -\frac{qe}{4\pi\varepsilon_0 a^3}x = m\frac{d^2x}{dt^2}.$$

We recognise this as the equation of simple harmonic motion, with angular frequency ω given by

$$\omega^2 = \frac{qe}{4\pi\varepsilon_0 a^3 m},$$

so using the relationship $\omega = 2\pi v$ gives

▶
$$v = \sqrt{\frac{eq}{16\pi^3 \varepsilon_0 a^3 m}}$$

as required.

Problem 124

An infinite line of charge λ per unit length is parallel to the line of intersection of two infinite conducting planes set at right angles to one another, such that it is a distance a from one and b from the other. Show that this arrangement is equivalent to four line charges so far as the electric field in the quadrant containing the original line charge is concerned. Hence calculate the charge induced per unit length on each plate.

Solution

This problem is most easily solved by the method of images, in which the electric field due to a charge near a conductor is calculated by introducing 'image charges' which, in combination with the original charge, produce a field which satisfies the boundary condition that the resultant field is normal to the surface of the conductor. For a plane conductor, the image charge is equal in magnitude and opposite in sign to the real charge, and is located in the 'mirror image' position on the other side of the conductor.

The geometry of the problem is shown in figure 107. In order to satisfy the boundary condition on the field at the horizontal plane, we can introduce an image line charge $-\lambda$ a distance a below it, and in order to satisfy the boundary condition at the vertical plane each of these line charges (real and image) must have an image to the left of the vertical plane. Thus the arrangement of line charges shown in figure 108 will produce the correct distribution of fields in the upper right-hand quadrant.

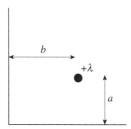

Figure 107

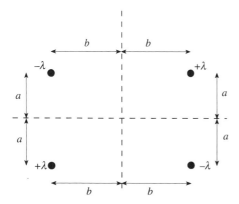

Figure 108

In order to calculate the resulting distribution of charge on the plates, let us first consider the effect of a single pair of line charges, as shown in figure 109.

From Gauss's theorem, the field E_1 at a distance r from a line charge λ per unit length is $\lambda/2\pi\varepsilon_0 r$, so the resultant field at x is

$$\frac{\lambda}{2\pi\varepsilon_0 r} 2\sin\theta = \frac{\lambda D}{\pi\varepsilon_0 r^2} \text{ downwards.}$$

Problem 125

A telephone wire of diameter 1 mm is suspended parallel to the ground at a height of 10 m. What is the capacitance to ground of this wire per unit length? (Assume the ground to be a conducting plane.)

Solution

The electric field distribution between the wire and the ground can be found by the method of images.

The real situation is as shown on the left of figure 112, but the field distribution between the wire and the ground will be reproduced by the arrangement shown on the right of the figure.

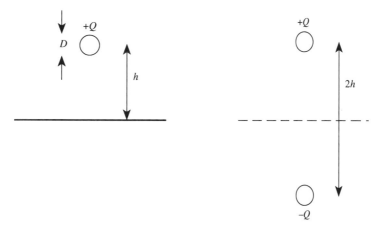

Figure 112

We saw in problem 118 that the potential difference V between two parallel wires of radius a separated by a distance d ($\gg a$) is

$$V = 2K \ln \frac{d}{a},$$

where

$$K = Q/2\pi\varepsilon_0,$$

one wire carrying a charge per unit length $+Q$ and the other carrying a charge per unit length $-Q$. By symmetry, the potential difference between one of the wires and the horizontal symmetry plane must be half of this value, so we can write the potential between the wire and the ground as

$$V_{\text{wg}} = K \ln \frac{d}{a} = \frac{Q}{2\pi\varepsilon_0} \ln \frac{d}{a} = \frac{Q}{2\pi\varepsilon_0} \ln \frac{2h}{(D/2)} = \frac{Q}{2\pi\varepsilon_0} \ln \frac{4h}{D},$$

where D is the diameter of the wire. The capacitance per unit length is thus

$$\frac{2\pi\varepsilon_0}{\ln \dfrac{4h}{D}}.$$

▶ Substituting $h = 10$ m and $D = 1$ mm gives 5.2×10^{-12} F m^{-1}.

Problem 126

Near the surface of the Earth there is a downward-directed electric field of 150 V m^{-1}. Using Gauss's theorem, calculate the surface charge density at the Earth's surface. Assume the Earth is a conducting medium.

At 200 m above the surface of the Earth the downward field is 100 V m^{-1}. Calculate the average volume charge density in the Earth's atmosphere below 200 m.

Express your answers as an excess (or deficit) of electrons per unit area or per unit volume.

Solution

Since the Earth is a conducting medium, there can be no electric field below the surface. Let us apply Gauss's theorem over a closed region of space having parallel, vertical sides, a horizontal upper surface of area A above the Earth's surface, and a horizontal lower surface (also of area A) below the Earth's surface, as shown in figure 113.

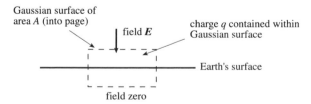

Gaussian surface of area A (into page)

field E

charge q contained within Gaussian surface

Earth's surface

field zero

Figure 113

Since the electric field is vertical, the sides of the region contribute nothing to $\int \mathbf{E} \cdot d\mathbf{s}$, and since the field is zero at the lower surface of the region, it also contributes zero to the integral. Thus $\int \mathbf{E} \cdot d\mathbf{s} = -EA$, where E is the magnitude of the electric field. (The minus sign arises from the fact that the vector area of the upper surface points upwards, whereas $\mathbf{E}$ points downwards.) The total charge contained within the region is therefore $-\varepsilon_0 EA$, so the charge per unit area is $-\varepsilon_0 E$. This corresponds to an excess of electrons at a density of $\varepsilon_0 E/e$ per unit area.
▶ Taking $E = 150 \text{ V m}^{-1}$ gives an excess of 8.3×10^9 electrons per m^2 at the Earth's surface.

200 m above the Earth's surface the downward field has fallen to 100 V m^{-1}. The apparent surface density of electrons as seen from a point at this altitude is thus $5.5 \times 10^9 \text{ m}^{-2}$, so the volume between zero metres and 200 metres must contain a deficit of electrons of $2.8 \times 10^9 \text{ m}^{-2}$. Assuming that the charge is uniformly distributed, this corresponds to a volume density of $(2.8 \times 10^9/200) \text{ m}^{-3}$, i.e. a deficit of 1.4×10^7 electrons
▶ per m^3.

[We can calculate this result directly by using the Maxwell equation

$$\mathbf{\nabla} \cdot \mathbf{E} = \frac{\rho}{\varepsilon_0},$$

where ρ is the charge density. Since $E_z = -150 \text{ V m}^{-1}$ when $z = 0$ and -100 V m^{-1} when $z = 200 \text{ m}$, $\partial E_z/\partial z = +0.25 \text{ V m}^{-2}$ assuming that ρ is constant. Since E_x and E_y are both zero, the value of $\mathbf{\nabla} \cdot \mathbf{E}$ is $+0.25 \text{ V m}^{-2}$, so $\rho = 2.2 \times 10^{-12} \text{ C m}^{-3}$. This corresponds to a deficit of electrons, with a volume density given by ρ/e as $1.4 \times 10^7 \text{ m}^{-3}$ as before.]

Problem 127

A parallel-plate capacitor is formed by two identical plates, each of area A, separated by a small distance d. One plate has a total charge Q and the other Q'. Neglecting edge effects, calculate the charge per unit area on each of the four metal surfaces and the electric field close to each surface.

By setting $Q' = -Q$, obtain an expression for the capacitance C.

Solution

Let us assume that a charge q is on the inside of the plate with total charge Q. By conservation of charge, the outside of this plate must carry a charge $Q - q$. Since the field between the plates is uniform (we are neglecting edge effects), the charge on the inside of the other plate must be $-q$, so by conservation of charge the outside of the plate must carry a charge $Q' + q$. Let us call the electric fields outside the capacitor E_1 and E_3, and the field between the plates E_2, as shown in figure 114.

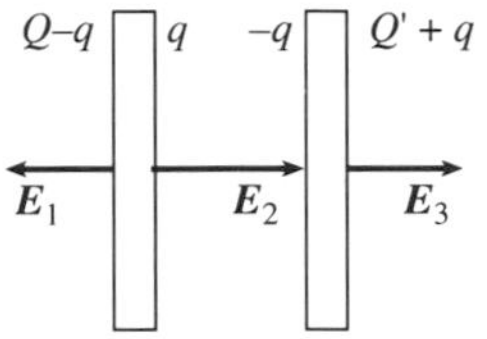

Figure 114

Gauss's theorem shows that if a charge q is uniformly distributed over a flat metal sheet of area A, the electric field in the direction away from the sheet is $q/\varepsilon_0 A$. Thus we have

$$E_1 = \frac{Q - q}{\varepsilon_0 A}, \ E_2 = \frac{q}{\varepsilon_0 A}, \ E_3 = \frac{Q' + q}{\varepsilon_0 A}.$$

In order to find the value of q, we need to consider the total energy of the system, which will be minimised at equilibrium. The energy per unit volume of an electric field E is

$$\tfrac{1}{2}\varepsilon_0 E^2.$$

It is clear that the volume occupied by the field E_2 is Ad. It is less obvious what volume is occupied by the fields E_1 and E_3. If the fields were truly uniform, the volumes would be infinite. In practice, the fields are not uniform because the field lines begin to diverge at large distances from the plates. We can allow for this by assuming that the fields are uniform with a large but finite volume v. The total energy U of the system can then be written as

$$U = \frac{1}{2\varepsilon_0 A^2}(v[Q - q]^2 + Ad\, q^2 + v[Q' + q]^2).$$

The condition that this is minimum is $dU/dq = 0$, so we differentiate U with respect to q to obtain

$$\frac{dU}{dq} = \frac{1}{\varepsilon_0 A^2}(v[2q - Q + Q'] + Ad\, q).$$

Setting this equal to zero, and rearranging, gives

$$q = \frac{Q - Q'}{2 + \dfrac{Ad}{v}}.$$

Now we can assume that $Ad/v \ll 1$, which gives $q = (Q - Q')/2$. [We can note in passing that this implies that the charges on the two outermost surfaces of the capacitor are equal.] Thus the charge per unit area on each surface, reading from left to right on the diagram, is

▶ $(Q + Q')/2A$,
    $(Q - Q')/2A$,
    $(Q' - Q)/2A$,
    $(Q + Q')/2A$,

and the fields are

▶ $E_1 = (Q + Q')/2\varepsilon_0 A$,
    $E_2 = (Q - Q')/2\varepsilon_0 A$,
    $E_3 = (Q + Q')/2\varepsilon_0 A$.

If $Q' = -Q$ (the normal arrangement for a capacitor), the external fields E_1 and E_3 are zero and the internal field $E_2 = Q/\varepsilon_0 A$. The potential difference across the gap d is thus

$$V = \frac{Qd}{\varepsilon_0 A},$$

▶   and since the capacitance C is defined by $Q = CV$, this gives $C = \varepsilon_0 A/d$.

Problem 128

A 15 nF capacitor is connected across a 70 V battery. How much work must be done in order to double the plate separation (a) with the battery connected, and (b) with it disconnected?

Solution

(a) Let us write C_0 for the initial capacitance of the capacitor, and V for the e.m.f. of the battery. Since we know that the capacitance of a parallel-plate capacitor is inversely proportional to the separation of the plates, the final capacitance must be $C_0/2$. The initial energy stored by the capacitor is $C_0 V^2/2$, and the final energy stored is $C_0 V^2/4$ if the battery is connected (so that the potential difference across the capacitor remains at V). Thus the energy stored by the capacitor has decreased by $C_0 V^2/4$. However, the charge stored in the capacitor must change from $C_0 V$ to $C_0 V/2$, so an amount of charge $C_0 V/2$ must be pushed 'backwards' through the battery, increasing its potential energy by $C_0 V^2/2$. The increase in the total energy of the system, and thus the work

which must be done on it, is thus $C_0V^2/4$. Substituting the values
$C_0 = 15$ nF and $V = 70$ V gives $18\ \mu\text{J}$.

(b) If the battery is disconnected, the charge on the capacitor must
remain constant while the capacitance is being changed. This will change
the potential difference across the capacitor. Since the relationship
between the charge Q, the capacitance C and the potential difference V
is $Q = CV$, halving the capacitance must double the potential difference.
Thus the energy stored by the capacitor after the plates have been
separated must be $\frac{1}{2}(C_0/2)(2V)^2 = C_0V^2$, i.e. an increase of $C_0V^2/2$.
Since the battery is disconnected this is the only energy term, so the work
done on the system must be $C_0V^2/2 = 37\ \mu\text{J}$.

Problem 129

An air-spaced parallel-plate capacitor has square plates of side l
separated by a distance t. Write down an expression for its capacitance C.

A square block of dielectric of side l, thickness t and relative
permittivity ε_r is now inserted so as to completely fill the space between
the plates. Calculate the change in the stored energy of the system:
  (i) if the plates have a constant charge Q,
  (ii) if a constant potential difference V is maintained between the
       plates by means of a battery.
With the battery still connected between the plates, the block of dielectric
is withdrawn in a direction parallel to one side of the plates until only a
length x remains between the plates (see figure 115). Find the magnitude
and direction of the force which acts on the block when it is in this
position. (Ignore edge effects throughout this problem.)

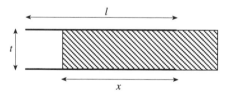

Figure 115

Solution

The capacitance C of the air-spaced capacitor is $\varepsilon_0 l^2/t$.

The capacitance C' of the capacitor when the gap is entirely filled with
dielectric of relative permittivity ε_r is $\varepsilon_0\varepsilon_r l^2/t$.

(i) The energy stored by a capacitor of capacitance C carrying a charge Q is $Q^2/2C$, so the energy change on inserting the dielectric when Q is held constant must be

▶
$$\Delta U - \frac{Q^2}{2}\left(\frac{1}{C'} - \frac{1}{C}\right) = \frac{Q^2 t}{2\varepsilon_0 l^2}\left(\frac{1}{\varepsilon_r} - 1\right).$$

(ii) The energy stored by a capacitor of capacitance C charged to a potential difference V is $V^2 C/2$, so the change in this energy on inserting the dielectric when V is held constant must be

$$\frac{V^2}{2}(C' - C) = \frac{V^2 \varepsilon_0 l^2}{2t}(\varepsilon_r - 1).$$

However, the charge stored on the capacitor increases by $(C' - C)V$ so the energy stored by the battery must decrease by $(C' - C)V^2$. The change in the total energy stored by the system is thus $-(C' - C)V^2/2$, so we can write

▶
$$\Delta U = -\frac{V^2 \varepsilon_0 l^2}{2t}(\varepsilon_r - 1).$$

To find the force on the dielectric slab when it is partially inserted, it is simplest to find an expression for the total energy of the system and then to differentiate it. The system consists, in effect, of two capacitors connected in parallel across a battery of potential V. The first of these capacitors is air-spaced and has an area of $l(l - x)$ and a separation t, so its capacitance is

$$C_1 = \varepsilon_0 l(l - x)/t.$$

The energy stored in its electric field is thus

$$C_1 V^2/2 = \varepsilon_0 l(l - x)V^2/2t.$$

The second capacitor is filled with the dielectric medium and has an area of lx and a separation t, so its capacitance is

$$C_2 = \varepsilon_0 \varepsilon_r lx/t.$$

The energy stored in its electric field is

$$C_2 V^2/2 = \varepsilon_0 \varepsilon_r lx V^2/2t,$$

so the total energy stored in the electric fields is

$$\frac{\varepsilon_0 l V^2}{2t}(l - x + \varepsilon_r x).$$

As before, however, we must also consider the energy stored in the battery. The total charge withdrawn from the battery is $C_1 V + C_2 V$, so the energy stored in it has *decreased* by $(C_1 + C_2)V^2$, which is numerically twice the value of the energy stored in the electric fields. The total energy of the system can thus be written as

$$U = -\frac{\varepsilon_0 l V^2}{2t}(l - x + \varepsilon_r x).$$

Differentiating this with respect to x gives

$$\frac{dU}{dx} = -\frac{\varepsilon_0 l V^2}{2t}(\varepsilon_r - 1),$$

▶ so the magnitude of the force is $\varepsilon_0(\varepsilon_r - 1)lV^2/2t$. Since U decreases as x increases, the force must act in the direction tending to increase x. The
▶ force is thus inwards, i.e. tending to move the dielectric further between the plates.

Problem 130

Estimate the capacitance of a thundercloud. If the breakdown electric field in air is $3 \times 10^6 \, \text{V m}^{-1}$, what charge flows down a lightning bolt?

Solution

A reasonable guess for the dimensions of a thundercloud might be a 1-km cube. If all the charge is separated to opposite faces of the cloud, it behaves like a parallel-plate capacitor with an area of $10^6 \, \text{m}^2$ and a separation of $10^3 \, \text{m}$. The formula $C = \varepsilon_0 A/d$ gives the capacitance as $8.85 \times 10^{-9} \, \text{F}$, but this precision is clearly not justified so we could
▶ estimate the capacitance as $\approx 10^{-8} \, \text{F}$.

The charge stored by a capacitor is given by CV, where V is the potential difference across the plates. If the field is $3 \times 10^6 \, \text{V m}^{-1}$ and the separation of the 'plates' is $10^3 \, \text{m}$, the potential difference is $3 \times 10^9 \, \text{V}$,
▶ giving a stored charge of about 30 C.

[A lightning discharge takes typically 1 ms, so the current is about 30 kA.]

Problem 131

An air-filled coaxial cable consists of a metal wire of diameter d surrounded by a thin metal sheath of diameter D. When the wire is charged, deduce expressions for the electric field $E(r)$ and electric potential $V(r)$ at a radial distance r ($d/2 < r < D/2$) from the central axis. Hence deduce the capacitance per unit length of the cable.

The breakdown field strength of air is $3\,\text{MV}\,\text{m}^{-1}$. If $d = 1$ mm and $D = 1$ cm, calculate the potential difference between wire and sheath at which electrical breakdown occurs.

Solution

By symmetry, the electric field is directed radially and depends only on distance from the axis. Let us assume that the inner wire carries a charge per unit length σ, and use Gauss's theorem to calculate the electric field E.

If we construct a Gaussian surface which is a cylinder of radius r and length l, coaxial with the capacitor (as shown in figure 116), the charge contained within it is σl. The surface integral $\int \mathbf{E} \cdot d\mathbf{s}$ is $2\pi r l E$, since the field $\mathbf{E}$ is everywhere constant and normal to the curved surface. Thus, by Gauss's theorem, $2\pi r l E = \varepsilon_0^{-1}\sigma l$ so

▶ $$E(r) = \frac{\sigma}{2\pi\varepsilon_0 r}.$$

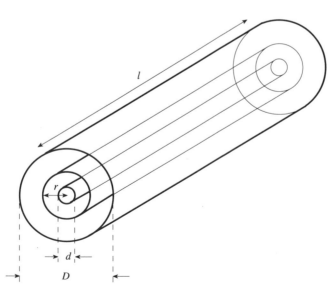

Figure 116

Now $\mathbf{E} = -\nabla V$, but since E is radial and varies only with r this simplifies to $E = -dV/dr$. Thus the potential V can be found by integrating $E(r)$ with respect to r. If we call the potential of the inner wire V_0, we have

$$V(r) - V_0 = -\frac{\sigma}{2\pi\varepsilon_0}\int_{d/2}^{r}\frac{1}{r}\,dr = -\frac{\sigma}{2\pi\varepsilon_0}\ln\left(\frac{2r}{d}\right).$$

Thus

$$V(r) = V_0 - \frac{\sigma}{2\pi\varepsilon_0}\ln\left(\frac{2r}{d}\right).$$

We can use this expression to evaluate the voltage between the capacitor's conductors by setting $r = D/2$, giving a potential difference of

$$V = \frac{\sigma}{2\pi\varepsilon_0}\ln\left(\frac{D}{d}\right).$$

Since the charge Q on the capacitor is σl, and the capacitance C is defined by $Q = CV$, the capacitance per unit length is

$$\frac{2\pi\varepsilon_0}{\ln(D/d)}.$$

The maximum electric field occurs where r is minimum, i.e. at $r = d/2$. If we set the field at $r = d/2$ equal to the breakdown field E_b, our expression for $E(r)$ shows that the charge per unit length σ on the capacitor must be $E_b\pi\varepsilon_0 d$. Substituting this value of σ into our expression for the potential difference V across the capacitor gives

$$V = \frac{E_b d}{2}\ln\left(\frac{D}{d}\right).$$

Taking $E_b = 3\times10^6\,\mathrm{V\,m^{-1}}$, $d = 1\,\mathrm{mm}$ and $D = 1\,\mathrm{cm}$ gives $V = 3.45\,\mathrm{kV}$.

Problem 132

A capacitor consists of two air-spaced concentric cylinders. The outer of radius b is fixed, and the inner is of radius a. If breakdown of air occurs at field strengths greater than E_b, show that the inner cylinder should have:

(a) radius $a = b/e$ if the potential of the inner cylinder is to be a maximum;

(b) radius $a = b/\sqrt{e}$ if the energy per unit length of the system is to be maximum, where e is the base of natural logarithms.

Solution

(a) From problem 131, we know that the potential difference V when the maximum field has its breakdown value is given by

$$V = E_b a \ln\left(\frac{b}{a}\right).$$

To find the maximum value of V if b is fixed, we differentiate this expression with respect to a and set $dV/da = 0$:

$$\frac{dV}{da} = E_b\left(\ln\left[\frac{b}{a}\right] - 1\right),$$

▶ so $dV/da = 0$ implies that $\ln(b/a) = 1$, therefore $b/a = e$ or $a = b/e$ as required.

(b) Again using a result from problem 131, we know that if the charge per unit length on the inner cylinder is σ, the field is

$$E = \frac{\sigma}{2\pi\varepsilon_0 r}.$$

The maximum field occurs at $r = a$, so if we set this maximum field to be E_b we can write

$$E = \frac{a E_b}{r}.$$

Now the energy per unit volume of an electric field E is $\varepsilon_0 E^2/2$, and the volume of a cylindrical shell of radius r, thickness dr and length L is $2\pi r L\, dr$, so the total energy U stored in a length L of the capacitor is

$$U = \frac{\varepsilon_0}{2} 2\pi L a^2 E_b^2 \int_a^b \frac{r\, dr}{r^2} = \varepsilon_0 \pi L a^2 E_b^2 \int_a^b \frac{dr}{r} = \varepsilon_0 \pi L a^2 E_b^2 \ln\left(\frac{b}{a}\right).$$

[We could also have derived this result using $U = CV^2/2$ and the expression for the capacitance obtained in problem 131.] Differentiating this with respect to a to find the maximum value of the energy per unit length gives

$$\frac{dU}{da} = \varepsilon_0 \pi L E_b^2\left(2a \ln\left[\frac{b}{a}\right] - a\right),$$

▶ so $dU/da = 0$ implies that $\ln(b/a) = 1/2$, therefore $b/a = \sqrt{e}$ or $a = b/\sqrt{e}$ as required.

Problem 133

A suggested use for a room-temperature superconductor is as an energy storage device. What is the maximum energy stored in a solenoid with the dimensions of a torch battery, length 0.05 m and diameter 0.03 m, if the maximum magnetic field sustainable by the superconductor is 15 T? How does this compare with the performance of a 1.5 V torch battery rated at 2 A-hours?

Solution

The energy per unit volume stored in a magnetic field B is

$$\frac{B^2}{2\mu_0},$$

so the energy stored in a cylindrical magnetic field B of length l and diameter D is

$$\frac{B^2}{2\mu_0} \frac{\pi D^2 l}{4}.$$

▶ Substituting the values gives an energy storage of 3.2 kJ.
   The torch battery can discharge a current of 2 A for 1 hour, so the
▶ energy stored is $1.5 \times 2 \times 3600 \, \text{J} = 10.8 \, \text{kJ}$.

Problem 134

A long metal rod of radius R carries a current I uniformly distributed over its cross section. Show on a sketch how the strength of the magnetic field B varies with distance r from the axis both inside and outside the rod.

Solution

By symmetry, the magnetic field must be azimuthal as shown in figure 117. The easiest way to calculate it is using Ampère's circuital theorem, which can be expressed as follows:

$$\int_{\text{loop}} \mathbf{B} \cdot d\mathbf{l} = \mu_0 I.$$

The line integral $\int \mathbf{B} \cdot d\mathbf{l}$ is performed round a closed loop, and I is the current enclosed by the loop.

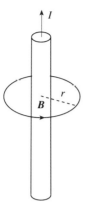

Figure 117

If we choose our loop to be a circle of radius r $(>R)$ centred on the rod's axis, it is clear that $\int \mathbf{B} \cdot d\mathbf{l} = 2\pi Br$. The current enclosed by the loop is I, so we have $B = \mu_0 I / 2\pi r$ for $r > R$.

For a point inside the rod $(r < R)$, we can again construct a circular loop centred on the rod's axis, and the loop integral will still be $2\pi Br$. However, the current encircled by the loop will be less in the ratio of the loop's area to the rod's cross section, i.e. it will be $I(r/R)^2$. Thus for $r < R$, we have $B = \mu_0 I r / 2\pi R^2$. We can thus sketch $B(r)$ as shown in figure 118.

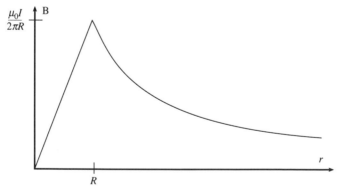

Figure 118

Problem 135

A coaxial cable consists of two thin coaxial cylinders electrically connected at one end; an inner cylindrical conducting tube of radius a

carrying a steady current I which is screened by an outer cylindrical conducting sheath of radius b which provides a return path. There is no dielectric medium present.

Using Ampère's theorem to derive the total magnetic energy stored in the space between the conductors, show that the inductance of a length l of the cable is

$$L = \frac{\mu_0 l}{2\pi} \ln\left(\frac{b}{a}\right).$$

If this cable ($a = 5$ mm, $b = 10$ mm, $l = 1000$ m) is now employed in a (resistanceless) LC circuit containing a capacitance $C = 1000\ \mu F$, determine the period of oscillations (neglect the capacitance of the cable itself).

Solution

We showed in problem 134 that the magnetic field at a radial distance r from the axis of a cylinder carrying a current I is

$$B = \frac{\mu_0 I}{2\pi r}.$$

Now the energy stored per unit volume of a magnetic field B is $B^2/2\mu_0$, so the total magnetic energy stored in a length l of this cable is

$$E = \int_a^b \left(\frac{\mu_0 I}{2\pi r}\right)^2 \frac{1}{2\mu_0} 2\pi r l\ dr = \frac{\mu_0 I^2 l}{4\pi} \int_a^b \frac{dr}{r} = \frac{\mu_0 I^2 l}{4\pi} \ln\left(\frac{b}{a}\right).$$

We also know that the magnetic energy stored by an inductance L carrying a current I is $LI^2/2$, so that equating these two expressions gives

▶ $$L = \frac{\mu_0 l}{2\pi} \ln\left(\frac{b}{a}\right)$$

as required.

Substituting the dimensions of the cable into this expression, we find that it has an inductance of 1.38×10^{-4} H. [In fact, this will be an underestimate of the inductance, because the magnetic fields within the conductors are not zero as we saw in problem 134. If we assume that the current in the inner conductor is uniformly distributed, and that the magnetic permeability of the conductor is unity, the field within it will add $\mu_0 l/8\pi$ to the inductance, which is not insignificant.]

We recall that the resonant frequency of an LC circuit is given by $\omega^2 = 1/LC$, so $\omega = 2.69 \times 10^3$ s^{-1} and the period of oscillation

▶ is $2\pi/\omega = 2.33$ ms.

[We can check that it is reasonable to neglect the capacitance of the cable itself by using the result from problem 131 that the capacitance per unit length is

$$\frac{2\pi\varepsilon_0}{\ln(b/a)}\cdot$$

Substituting the dimension of the cable into this expression gives a capacitance per unit length of $8.0 \times 10^{-11}\,\mathrm{F\,m}^{-1}$, so that the capacitance of a cable 1000 m long is $0.08\,\mu\mathrm{F}$, which is clearly negligible compared with the $1000\,\mu\mathrm{F}$ capacitor connected externally.]

Problem 136

A long straight copper wire, of circular cross section, contains n conduction electrons per unit volume, each of charge q. Show that the current I in the wire is given by

$$I = nqv\pi a^2,$$

where v is the drift velocity and a is the radius of the wire.

At a radial distance r from the axis of the wire, what is the direction of the magnetic field $\mathbf{B}$ due to the current I? Assuming that the magnitude of the field is $B = \mu_0 I/2\pi r$ $(r \geqslant a)$, obtain an expression for the Lorentz force $\mathbf{F}$ on an electron moving with the drift velocity at the surface of the wire.

If $I = 10$ A and $a = 0.5$ mm, calculate the magnitude of (a) the drift velocity and (b) the force, given that for copper, $n = 8.5 \times 10^{28}\,\mathrm{m}^{-3}$.

Solution

The current I in the wire is defined as the rate at which charge crosses a plane perpendicular to the direction of the current. Mathematically,

$$I = \frac{dQ}{dt} = \frac{dQ}{dx}\frac{dx}{dt},$$

i.e. the current is given by the product of the charge per unit length of the wire with the mean velocity of the charge carriers. Clearly, $dQ/dx = nqA$ where n is the number of charge carriers per unit volume, q is their charge and A is the cross-sectional area of the wire. Thus $I = nqv\pi a^2$ as required.

Using Fleming's right-hand rule, the direction of the field at X (see figure 119) is out of the page if the conventional current (flow of positive

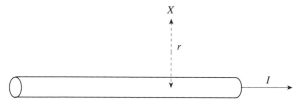

Figure 119

charge) is to the right as shown. The electron drift is to the left since the electrons are negatively charged. The Lorentz force **F** on a particle of charge q moving with velocity **v** in a magnetic field **B** is

$$\mathbf{F} = q\mathbf{v} \times \mathbf{B},$$

so the Lorentz force on an electron acts inwards. The magnitude of the magnetic force on an electron at the surface of the wire is

$$F = Bqv = \frac{\mu_0 I}{2\pi a} q \frac{I}{nq\pi a^2} = \frac{\mu_0 I^2}{2\pi^2 n a^3}.$$

If $I = 10$ A, $a = 0.5$ mm and $n = 8.5 \times 10^{28}$ m^{-3}, we find
▶ $v = 9.4 \times 10^{-4}$ m s^{-1} and $F = 6.0 \times 10^{-25}$ N.

Problem 137

Two coaxial plane coils, each of n turns of radius a, are separated by a distance a. Calculate the magnetic field on the axis at the point midway between them when a current I flows in the same sense through each coil.

   Electrons in a colour television tube are accelerated through a potential difference of 25 kV and then deflected by 45° in the magnetic field between the two coils described above. If a is 100 mm and the maximum current available for the coils is 2 A, estimate the number of turns which the coils must have.

Solution

To find the magnetic field on the axis of a circular coil, at a distance x from the plane of the coil, we can use the Biot–Savart law:

$$d\mathbf{B} = \frac{\mu_0 I}{4\pi} \frac{d\mathbf{l} \times \mathbf{r}}{r^3},$$

where $d\mathbf{B}$ is the contribution to the magnetic field from a current I flowing in an element of wire of vector length $d\mathbf{l}$ whose position vector is $\mathbf{r}$ relative to the point at which we are calculating the field (see figure 120).

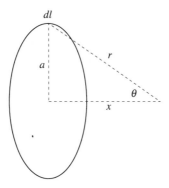

Figure 120

Since $d\mathbf{l}$ and $\mathbf{r}$ are mutually perpendicular, their cross product $d\mathbf{l} \times \mathbf{r}$ has magnitude $r\,dl$, and its direction must be in the plane of the triangle (with sides a, r and x) and perpendicular to the vector $\mathbf{r}$. By symmetry we can see that, when we integrate over the whole loop, only the components of $\mathbf{B}$ which lie along the axis will not cancel out. Thus we can write

$$dB = \frac{\mu_0 I\,dl}{4\pi r^2}\sin\theta.$$

Integrating round the whole coil involves l running from zero to $2\pi nl$ (since there are n turns each of length $2\pi a$), so the total field is

$$B = \frac{\mu_0 I 2\pi na}{4\pi r^2}\sin\theta = \frac{\mu_0 nI a^2}{2(a^2 + x^2)^{3/2}}.$$

The field at $x = a/2$ (midway between the coils) is thus

$$\frac{\mu_0 nI}{2a(5/4)^{3/2}},$$

so the field due to both coils together is

$$\left(\frac{4}{5}\right)^{3/2}\frac{\mu_0 nI}{a}.$$

The constant $(4/5)^{3/2}$ has a value of 0.716.

If we assume that the magnetic field strength between the deflecting coils has a constant value B, given by the expression we have just calculated, everywhere between the coils and is zero outside this region, the electrons will describe a circular path in the magnetic field, with an angular velocity $\omega = eB/m$.

When the electrons have travelled a distance $x = a$ (measured parallel to their original direction, starting from the point at which they enter the field), they have been deflected through an angle of $45° = \pi/4$ radians as shown in figure 121. The relationship between a and the radius r of the circular path is thus

$$a = r \sin (\pi/4) = r/\sqrt{2},$$

so

$$r = a\sqrt{2}.$$

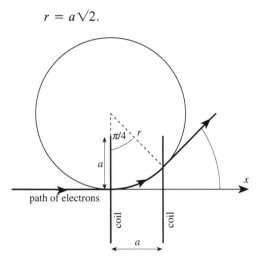

Figure 121

Now if the electrons are travelling at speed v (which is unchanged by the magnetic field), the relationship between v, r and ω is

$$v = r\omega,$$

so we must have

$$v = \frac{\sqrt{2}\, aeB}{m}.$$

Putting

$$v = \sqrt{\frac{2eV}{m}},$$

where V is the accelerating voltage, and

$$B = 0.716\frac{\mu_0 nI}{a},$$

and rearranging to make n the subject, gives

$$n = \frac{1.397}{\mu_0 I}\sqrt{\frac{mV}{e}}.$$

We can now substitute $I = 2\,A$ and $V = 25\,kV$ (the radius of the coils does not matter) to obtain $n = 210$. Since our assumption that the field is zero outside the coils is not a very accurate one, we can state the answer as $n \approx 200$.

[The arrangement of two equal coils separated by a distance equal to their radii is known as *Helmholtz coils*. It produces a field on the axis between the coils which is nearly uniform, as we have assumed. Using our expression for the magnetic field on the axis of a single coil, we can calculate the variation of the field due to two coils, as shown in figure 122.

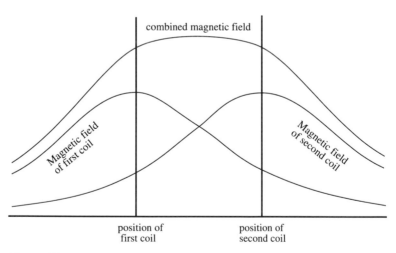

Figure 122

It can be seen that our assumption that the field is uniform between the coils is a very reasonable one. The assumption that the field is zero outside the region between the coils is less reasonable.]

Problem 138

Use Ampère's law to find the magnetic field strength well inside a very long solenoid of length l, cross-sectional area A and total number of turns N, carrying current I. Show that the magnetic field strength at the centre of the solenoid is approximately twice the value at each end, on the axis of the solenoid.

If the solenoid is long enough for these 'end effects' to be completely neglected, calculate its self-inductance L. How much work is done in establishing a current I?

Solution

Consider a side view of the solenoid, as shown in figure 123. Let us assume that the magnetic field inside the solenoid has a magnitude B and a direction which is parallel to the axis, and that the magnetic field outside the solenoid is zero. If we construct a loop whose length is x in the direction of the solenoid axis, as shown by the dashed line, the loop will enclose Nx/l turns and therefore encircle a current of NIx/l. The line integral $\int \mathbf{B} \cdot d\mathbf{l}$ will have a value of Bx, so Ampère's theorem gives

$$Bx = \frac{\mu_0 NIx}{l}$$

▶ and thus $B = \mu_0 NI/l$.

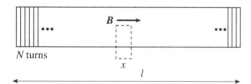

N turns

x

l

Figure 123

To find the magnetic field strength at the ends of the solenoid, on the axis, we can use the Biot–Savart law to find the field at a general position inside the solenoid, which we will assume to have a circular cross-section of radius a. Let us consider the field at a distance x from an element of the solenoid of length dx, as shown in figure 124.

The current carried by the element is clearly $NI\,dx/l$, so the magnetic field at x has a direction along the solenoid axis (by symmetry) and a magnitude dB equal to

$$\frac{\mu_0 NI\,dx}{4\pi l}\,\frac{2\pi a}{r^2}\sin\theta.$$

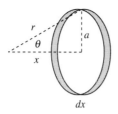

Figure 124

We can write this in terms of a and θ by using

$$r = a/\sin\theta,$$
$$x = a/\tan\theta,$$

so

$$dx = -a\, d\theta/\sin^2\theta.$$

On substitution of these values, we obtain

$$dB = -\frac{\mu_0 NI}{2l}\sin\theta\, d\theta.$$

This can now be integrated to give the total field B in the solenoid:

$$B = \frac{\mu_0 NI}{2l}(\cos\theta_1 - \cos\theta_2),$$

where θ_1 and θ_2 are the values of θ at the two ends of the solenoid, as seen from the point at which we wish to determine the field strength, as shown in figure 125.

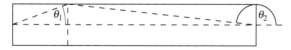

Figure 125

For a point well inside a very long solenoid, θ clearly ranges from 0 to π, which gives the result $B = \mu_0 NI/l$ that we obtained earlier. For a point at one end of the solenoid, θ ranges from $\pi/2$ to π, which gives a value half as large.

[There is a much easier way of deriving this result for the field strength at the ends of the solenoid, although it will not give us an expression for B at a general point. We note that the expression $B = \mu_0 NI/l$ for the field well inside a long solenoid is proportional to the number of turns per unit length. If we take an infinitely long solenoid and chop it into two halves, we will produce two solenoids which are essentially identical to the original one, i.e. still infinite in length, and, since they will have the same number of turns per unit length, the distribution of the magnetic field inside them will be the same as in the original solenoid. Since the middle of the original solenoid is converted into two ends by this process, we can see that the field strength at the middle must be equal to twice the field strength at the ends.]

If the solenoid is long enough to neglect the 'end effects' (i.e. much longer than its diameter so that we can assume the magnetic field to be constant along the solenoid), the total flux linked is

$$\Phi = BNA = \frac{\mu_0 N^2 A I}{l},$$

and since $\Phi = LI$ where L is the self-inductance, we obtain

▶ $L = \mu_0 N^2 A / l$.

To find the work done in establishing a current I in the inductor, we can use the result that the energy per unit volume of a magnetic field B is $B^2/2\mu_0$ to calculate the total energy U as

$$U = \frac{1}{2\mu_0}\left(\frac{\mu_0 NI}{l}\right)^2 Al = \frac{\mu_0 N^2 A I^2}{2l}.$$

▶ This can be rewritten in terms of the self-inductance L as $U = LI^2/2$.

[We can derive this result more generally as follows: If the current in the inductor is changing, the induced e.m.f. $\mathscr{E}$ is given by

$$\mathscr{E} = -\frac{d\Phi}{dt} = -L\frac{dI}{dt}.$$

The rate of doing work on the system, dW/dt, is given by $-\mathscr{E}I$, so we have

$$\frac{dW}{dt} = LI\frac{dI}{dt} = \frac{1}{2}\frac{d}{dt}(LI^2).$$

Integrating from $I = 0$ to some current I, we thus find that $W = LI^2/2$. Since $U = 0$ when $I = 0$, the work done on the system, W, must be equal to the total energy of the magnetic field.]

Problem 139

A metallic ring of cross-section 2.5 cm², mean radius 40 cm and relative permeability 1500 is wound uniformly with 3000 turns of wire. If a current of 1.6 A passes through the wire, find the mean **B** field and magnetisation in the ring.

Solution

In the presence of a magnetic medium, Ampère's circuital theorem becomes

$$\int_{\text{loop}} \mathbf{H} \cdot d\mathbf{l} = I,$$

where **H** is the magnetic intensity. The magnetic intensity is related to the magnetic flux density **B** and the magnetisation (magnetic dipole per unit volume) **M** by

$$\mathbf{H} = \frac{\mathbf{B}}{\mu_0} - \mathbf{M}$$

and, in an isotropic medium, **B** and **H** are related by

$$\mathbf{B} = \mu\mu_0\mathbf{H},$$

where μ is the relative permeability. Combining these two expressions, we can write **M** in terms of **H** as

$$\mathbf{M} = (\mu - 1)\mathbf{H}.$$

If we put N for the number of turns of wire and I for the current flowing in it, it is clear that the total current encircling the ring is NI. Ampère's circuital theorem shows that this is equal to $2\pi r H$ where r is the radius of the ring, so we must have

$$H = \frac{NI}{2\pi r}.$$

The mean **B** field is thus

$$\frac{\mu\mu_0 NI}{2\pi r}.$$

Substituting $\mu = 1500$, $N = 3000$, $I = 1.6$ A and $r = 0.40$ m gives
▶ $B = 3.6$ T. The mean magnetisation **M** is

$$\frac{(\mu - 1)NI}{2\pi r}$$

▶ and substitution of the values gives $M = 2.9 \times 10^6$ A m^{-1}. [The area of the ring is not needed in the solution.]

Problem 140

The anchor-ring specimen of ferromagnetic material shown in figure 126 has a relation $B(H)$ of the form

$$B^2 + \mu_0^2 H^2 = K^2$$

for fields such that **B** and **H** are antiparallel, where K is a constant.

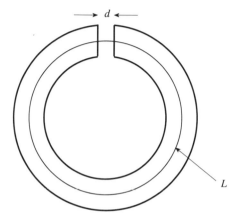

Figure 126

The circumference is L and there is a narrow air-gap of width d. Obtain an expression for the magnetic flux density in the air-gap in terms of the constants d, L and K.

Solution

In the presence of magnetic media, Ampère's circuital theorem becomes

$$\int_{\text{loop}} \mathbf{H} \cdot d\mathbf{l} = I,$$

and since no current passes round the ring, this integral must be zero. Thus we may write

$$H_{\text{ring}}(L - d) + H_{\text{gap}}d = 0.$$

However, in free space (to which air is a very good approximation) $B = \mu_0 H$, so the magnetic flux density in the air gap is given by $\mu_0 H_{\text{gap}}$. This field is perpendicular to the plane surfaces of the ring, and the perpendicular component of the **B** field is constant at an interface, so B is constant throughout the ring. We can thus abandon the subscripts, and write

$$H(L - d) + \frac{Bd}{\mu_0} = 0,$$

where H is the magnetic intensity in the ring and B is the flux density in the ring and in the gap.

The solution we are looking for is the value of B which simultaneously satisfies this relation and the $B(H)$ relation for the material of the ring. A graph of the $B(H)$ relation has the form of part of an ellipse, or if we change the scales suitably, a circle (see figure 127).

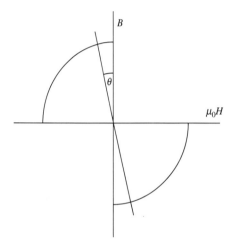

Figure 127

The circle has radius K, and since the given relation between B and H applies only when the fields are antiparallel, only those two quadrants have been drawn. The straight line represents the relation

$$H(L - d) + \frac{Bd}{\mu_0} = 0$$

and it makes an angle θ with the B-axis where

$$\tan \theta = \frac{d}{L - d}.$$

By inspection of the diagram, we can see that the value of B must be $\pm K \cos \theta$. Since

$$\cos^2 \theta = \frac{1}{1 + \tan^2 \theta},$$

we have

$$B^2 = \frac{K^2}{1 + \left(\dfrac{d}{L-d}\right)^2} = \frac{K^2(L-d)^2}{(L-d)^2 + d^2}.$$

Thus

▶ $$B = \pm K \frac{L-d}{\sqrt{([L-d]^2 + d^2)}}.$$

[For $d \ll L$, we can use the binomial expansion to approximate this as $\pm K(1 - d^2/2L^2)$.]

Problem 141

A coil of wire consisting of twenty turns in the shape of an equilateral triangle of side 5 cm is placed in a magnetic field of 10^{-3} T parallel to the plane of the triangle. What is the torque acting on the coil when a current of 0.1 A is allowed to flow?

Solution

The easiest way to answer this is to recall that when the coil is carrying a current it has a magnetic moment $\mathbf{M}$, and that the torque $\mathbf{G}$ on a magnetic moment $\mathbf{M}$ in a magnetic field $\mathbf{B}$ is given by $\mathbf{G} = \mathbf{B} \times \mathbf{M}$.

The magnetic moment of a coil is given by $NI\mathbf{A}$, where N is the number of turns and $\mathbf{A}$ is the vector area of the coil. In this case the coil is a twenty-turn equilateral triangle of side a, so the magnitude of the magnetic moment is

$$M = NI \frac{a^2\sqrt{3}}{4}.$$

The direction of $\mathbf{M}$ is normal to the plane of the triangle, and therefore normal to the direction of $\mathbf{B}$, so the torque is numerically equal to BM.
▶ Substituting the values given in the problem gives $G = 2.2 \times 10^{-6}$ N m.

Problem 142

Two horizontal metal rails, separated by a distance L, run parallel to the x-axis. At $x = 0$ a resistor R is connected between the rails. A closed

(a) If the spectrometer were used with $100\ \mathrm{V\,cm^{-1}}$ between the plates and a magnetic field of 0.2 T, what would be the speed of an ion that can pass through the velocity filter?

(b) If the velocity-filter exit slit were 1 mm wide, could this machine resolve the two isotopes?

Solution

Figure 130 shows the arrangement of a Bainbridge mass spectrometer.

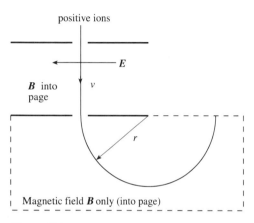

positive ions

E

B into page

v

r

Magnetic field B only (into page)

Figure 130

(a) Within the velocity selector (the region containing both E and B fields), a positive ion of charge q and speed v will experience a force Eq to the left as a result of the electric field, and a force Bqv to the right as a result of the magnetic field. The ions that travel in a straight line (and so emerge from the velocity selector) must therefore have $v = E/B$. Taking $E = 10^4\ \mathrm{V\,m^{-1}}$ and $B = 0.2\ \mathrm{T}$ gives $v = 5 \times 10^4\ \mathrm{m\,s^{-1}}$.

(b) Within the region containing only a magnetic field, the ions will be deflected into a circular arc of radius r. Equating the magnetic force Bqv to the centripetal force mv^2/r gives the radius of the arc as

$$r = \frac{mv}{Bq}$$

and clearly the sideways deflexion of an ion is given by $2r$. The deflexion of a singly charged ^{3}He ion is thus

$$\frac{2 \times 3 \times 1.66 \times 10^{-27} \times 5 \times 10^4}{0.2 \times 1.60 \times 10^{-19}}\mathrm{m} = 15.6\ \mathrm{mm}$$

and the deflexion of a singly charged ^{4}He ion will be 4/3 times as large, or 20.8 mm. The separation between the ends of the two trajectories will thus be 5.2 mm. The exit slit width of 1 mm will broaden the two trajectories into bands of width 1 mm, so they will not overlap and the isotopes will be resolved.

Problem 148

Two long concentric cylindrical conductors of radii a and b ($b < a$) are maintained at a potential difference V and carry equal and opposite currents I. Show that an electron with a particular velocity u parallel to the axis may travel undeviated in the evacuated region between the conductors, and calculate u when $a = 50$ mm, $b = 2.0$ mm, $V = 50$ V and $I = 100$ A.

It is also possible for the electron to travel in a helical path. By regarding such a path as the combination of a circular motion perpendicular to the axis with a steady velocity parallel to the axis, indicate without detailed mathematics how this comes about.

Solution

From problem 131, we know that the electric field E in the region between the conductors varies with the distance r from the axis as

$$E = \frac{\sigma}{2\pi\varepsilon_0 r},$$

where σ is the charge per unit length on the inner cylinder, and we also know that the potential difference V between the conductors is given by

$$V = \frac{\sigma}{2\pi\varepsilon_0} \ln\left(\frac{a}{b}\right).$$

Combining these expressions to eliminate σ gives

$$E = \frac{V}{r \ln\left(\dfrac{a}{b}\right)}.$$

From problem 134 we know that the magnetic field B varies with r as

$$B = \frac{\mu_0 I}{2\pi r}.$$

the dashed curve gives

$$B = \frac{\mu_0 I}{b}.$$

To calculate the flux linked by the coil we must multiply this flux density by the area of the coil measured perpendicularly to the direction of **B**. This area is clearly la, so the flux linked by the circuit is

$$\Phi = \frac{\mu_0 I l a}{b},$$

▶   and the self-inductance of the coil is given by Φ/I, so $L = \mu_0 la/b$.

The energy U stored by the magnetic field is given, in general, by $U = LI^2/2$. Substituting our expression for the self-inductance L gives

▶ $$U = \frac{\mu_0 la \, I^2}{2b}.$$

(a) If the self-inductance is changed without changing the current, the change in the energy stored by the magnetic field is clearly

▶ $$\delta U = \frac{\mu_0 I^2 l}{2b} \delta a.$$

(b) If the circuit contains only the coil, no e.m.f. is supplied so there can be no change in the flux linked by the coil, and so the current will change as the inductance changes. We can rewrite the energy stored in terms of the flux linked as

$$U = \frac{\Phi^2 b}{2\mu_0 la},$$

so the change in U when a is changed by δa is given by

$$\delta U = -\frac{\Phi^2 b}{2\mu_0 la^2} \delta a.$$

Rewriting this in terms of the current I gives

▶ $$\delta U = -\frac{\mu_0 I^2 l}{2b} \delta a.$$

Since this is negative, the change is energetically favourable so the forces on the plates must be outwards (repulsive). The force F is $-dU/da$, so the force per unit area on each plate is

▶ $$\frac{\mu_0 I^2 l}{2b} \cdot \frac{1}{lb} = \frac{\mu_0 I^2}{2b^2}.$$

Problem 150

A circular parallel-plate capacitor of radius a and plate separation d is connected in series with a resistor R and a switch, initially open, to a constant voltage source V_0. The switch is closed at time $t = 0$. Assuming that the charging time of the capacitor, $\tau = CR$, is very long compared with a/c and that $d \ll a$ (C is the capacitance and c is the velocity of light), find an expression for the displacement current density as a function of time. Obtain an expression for the magnetic flux density **B** as a function of time and of position between the capacitor plates.

Solution

The charge on the capacitor plates at time t is given by

$$Q = CV_0(1 - \exp[-t/\tau]),$$

so the charge density on the plates is

$$\frac{CV_0}{A}(1 - \exp[-t/\tau]),$$

where $A = \pi a^2$ is the area of the plates. By Gauss's theorem the charge density is equal to $\varepsilon_0 E$, where E is the electric field between the plates, and since the displacement current density j_d is equal to the rate of change of $\varepsilon_0 E$ we must have

$$j_d = \frac{CV_0}{A\tau} \exp(-t/\tau).$$

Since $\tau = CR$ and $A = \pi a^2$, we may write this as

▶ $$j_d = \frac{V_0}{\pi a^2 R} \exp(-t/\tau).$$

To find the magnetic flux density B between the plates we can apply Ampère's circuital theorem.

If we consider a circular loop of radius r concentric with the capacitor, as shown in figure 134, it links a total displacement current

$$I_d = \pi r^2 j_d = \frac{r^2 V_0}{a^2 R} \exp(-t/\tau).$$

Ampère's theorem shows that $\mu_0 I_d$ is equal to the loop integral $\int \mathbf{B} \cdot d\mathbf{l}$ round the loop (since there are no free currents between the capacitor

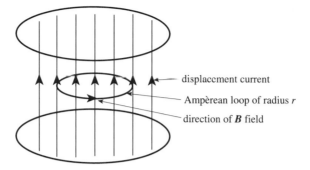

displacement current

Ampèrean loop of radius r

direction of $\mathbf{B}$ field

Figure 134

plates). Since $\mathbf{B}$ has azimuthal symmetry, the integral is just $2\pi r B$, so we obtain

$$B = \frac{\mu_0 r V_0}{2\pi a^2 R} \exp(-t/\tau).$$

[The condition that $\tau \gg a/c$ implies that the energy stored in this magnetic field can be ignored in comparison with the energy stored in the electric field between the plates. We can see this as follows. The maximum energy E_e stored in the electric field (when the capacitor is fully charged) is $CV_0^2/2$. The maximum energy E_m stored in the magnetic field can be found by integrating the energy per unit volume $B^2/2\mu_0$ over volume at time $t = 0$:

$$E_m = \frac{1}{2\mu_0}\left(\frac{\mu_0 V_0}{2\pi a^2 R}\right)^2 d\int_0^a r^2 2\pi r\, dr = \frac{1}{2\mu_0}\left(\frac{\mu_0 V_0}{2\pi a^2 R}\right)^2 \frac{d \cdot 2\pi \cdot a^4}{4}.$$

The ratio of these terms is

$$\frac{E_m}{E_e} = \frac{\mu_0 d}{8\pi R^2 C}.$$

Now we can write $R = \tau/C \gg a/Cc$, so

$$\frac{E_m}{E_e} \ll \frac{\mu_0 d C c^2}{8\pi a^2}.$$

Putting $C \simeq \varepsilon_0 \pi a^2/d$ and $\varepsilon_0 \mu_0 = 1/c^2$ gives $E_m/E_e \ll 1/8$.]

Electric circuits

Problem 151

A current of 1 mA enters the resistor network shown in figure 135 at A
and leaves at B. Find the current in each resistor.

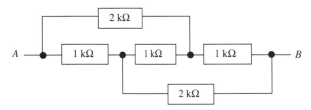

Figure 135

Solution

For convenience, we will work with the numerical values of currents,
resistances and voltages, in mA, kΩ and V respectively to give a
consistent set of units. Let us label the two internal connexions C and D,
as shown in figure 136, and denote the current flowing in the upper 2 kΩ
resistor by a. Since a current of 1 enters the network at A, the current in
the 1 kΩ resistor between A and C must, by Kirchhoff's first law, be
$1 - a$. Let us also denote the current in the resistor between C and D by
b. Again applying Kirchhoff's first law, we can see that the current in the
lower 2 kΩ resistor must be $1 - a - b$ and that the current in the resistor
between D and B must be $a + b$.

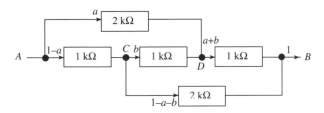

Figure 136

273

Now we can apply Kirchhoff's second law by considering the voltages dropped across the various resistors. Since the current in the upper $2\,k\Omega$ resistor is a, the voltage dropped across it is $2a$. Similarly, the voltage dropped across the resistor between A and C is $1 - a$ and the voltage dropped across the resistor between C and D is b. Thus the voltage drop from A to D is $1 - a + b$, which must be equal to $2a$, so we have

$$1 - a + b = 2a.$$

The voltage drop across the lower $2\,k\Omega$ resistor is $2(1 - a - b)$, and the voltage drop across the resistor between D and B is $a + b$, so we can write the following equation for the voltage drop from C to B:

$$2(1 - a - b) = b + a + b = a + 2b.$$

Solving these two simultaneous equations gives $a = 2/5$ and $b = 1/5$, so we can now write down the currents as follows:

▶   Upper $2\,k\Omega$ resistor: $a = \frac{2}{5}\,mA.$

▶   Left-hand $1\,k\Omega$ resistor: $1 - a = \frac{3}{5}\,mA.$

▶   Middle $1\,k\Omega$ resistor: $b = \frac{1}{5}\,mA.$

▶   Right-hand $1\,k\Omega$ resistor: $a + b = \frac{3}{5}\,mA.$

▶   Lower $2\,k\Omega$ resistor: $1 - a - b = \frac{2}{5}\,mA.$

Problem 152

Find the Thévenin equivalent for the circuit to the left of terminals A and B in figure 137. Calculate the current in and the voltage across the $15\,\Omega$ load resistance.

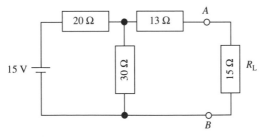

Figure 137

If R_L were a variable resistance, what should be its value for maximum power to be developed in R_L? What is the value of this maximum power?

Solution

A Thévenin equivalent circuit consists of an e.m.f. E in series with a resistance R, with E and R chosen such that the output from the circuit will be the same as the output appearing at the terminals A and B under the same loading conditions.

If we first consider the effect of removing the load resistance R_L, as shown in figure 138, we can see that the voltage across the terminals of the Thévenin circuit must be E. The 20 Ω and 30 Ω resistors of the original circuit act as a potential divider, so that the voltage across AB is

$$15 \frac{30}{30 + 20} \text{ V} = 9 \text{ V}.$$

▶ Thus $E = 9$ V.

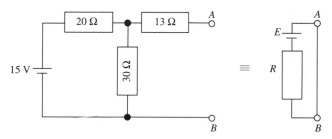

Figure 138

Now we consider the effect of connecting A and B with a short-circuit as shown in figure 139. It is clear that the current flowing between A and B in the Thévenin circuit must be E/R. In the original circuit, the connexion has the effect of connecting the 13 Ω and 30 Ω resistors in parallel, giving a combined resistance of 9.07 Ω. The total resistance into which the battery is discharging is thus 29.07 Ω, so the total current drawn from the battery is 0.5160 A. This current is divided, with a fraction 30/43

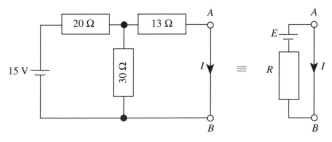

Figure 139

passing through the 13 Ω resistor and hence from A to B. The current flowing between A and B is thus 0.3600 A. Equating this to ► E/R and taking $E = 9$ V gives $R = 25\,\Omega$.

When the load resistance has a value of 15 Ω, the Thévenin equivalent circuit consists of a 9 V battery driving a current through a total resistance ► of 40 Ω, so the current is $9/40 = 0.225$ A. The voltage across the load is ► $15 \times 0.225 = 3.375$ V.

If the load resistance is variable, the total resistance is $R + R_L$ so the current is

$$\frac{E}{R + R_L}$$

and the power P dissipated in the load is

$$P = \left(\frac{E}{R + R_L}\right)^2 R_L.$$

To find the value of R_L for which P is maximum we set $dP/dR_L = 0$:

$$\frac{dP}{dR_L} = \frac{E^2}{(R + R_L)^2} - \frac{2E^2 R_L}{(R + R_L)^3} = 0 \quad \therefore 2R_L = (R + R_L).$$

► Thus the appropriate value of R_L is $R = 25\,\Omega$. (This is the maximum power theorem.)

Substituting this value into our expression for P gives
► $P = E^2/4R = 0.81$ W.

Problem 153

The characteristics of an electrical device are examined by measuring the voltage drop across various load resistors connected between its terminals, as shown in figure 140.

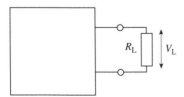

Figure 140

When $R_L = 1\,\text{k}\Omega$, $V_L = 75\,\text{mV}$.
When $R_L = 100\,\text{k}\Omega$, $V_L = 5\,\text{V}$.

Deduce from these data the Thévenin equivalent circuit representing the device and calculate the voltages produced across loads of 1 MΩ and 10 MΩ.

Suppose that the output of the device varies sinusoidally at a frequency of 1 Hz and that there is a large 50 Hz noise signal due to mains pickup. Show how a capacitor could be added to the circuit to filter out the unwanted 50 Hz signal. For a load R_L of infinite resistance, calculate the value of the capacitance required to attenuate the 1 Hz signal by 3 dB. With this capacitor, calculate by what factor the 50 Hz noise is attenuated and express this attenuation in dB.

Solution

Replacing the device by its Thévenin equivalent (having an e.m.f. E and a resistance R) gives the circuit shown in figure 141.

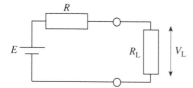

Figure 141

The two resistors act as a potential divider, and V_L is given by

$$V_L = \frac{ER_L}{R + R_L}.$$

Taking the reciprocal of this expression (for convenience) and rearranging it gives

$$\frac{1}{E} = \frac{1}{V_L} - \frac{R}{E}\frac{1}{R_L} = \text{constant.} \tag{1}$$

Thus if we put R_1 and R_2 for the two values of R_L, and V_1 and V_2 for the corresponding values of V_L, we obtain

$$\frac{1}{V_1} - \frac{R}{E}\frac{1}{R_1} = \frac{1}{V_2} - \frac{R}{E}\frac{1}{R_2}. \tag{2}$$

This can be rearranged as

$$\frac{R}{E} = \frac{1/V_1 - 1/V_2}{1/R_1 - 1/R_2}. \tag{3}$$

Substituting (3) into (1) gives

$$E = \frac{R_2 - R_1}{R_2/V_2 - R_1/V_1} \tag{4}$$

and substituting (4) into (3) gives

$$R = \frac{V_2 - V_1}{V_1/R_1 - V_2/R_2}. \tag{5}$$

Inserting the values $R_1 = 10^3\ \Omega$, $R_2 = 10^5\ \Omega$, $V_1 = 75 \times 10^{-3}$ V and $V_2 = 5$ V into equation (4) gives $E = 14.85$ V. Inserting the values into (5) gives $R = 197$ kΩ.

Across a load resistor of 1 MΩ, the voltage would be

$$14.85 \times 10^6/(197 \times 10^3 + 10^6) = 12.41 \text{ V}.$$

Across 10 MΩ, the voltage would be

$$14.85 \times 10^7/(197 \times 10^3 + 10^7) = 14.56 \text{ V}.$$

In order to filter out high frequencies (where by 'high' we mean much larger than 1 Hz), we should connect a capacitor across the output terminals of the device, as shown in figure 142. The impedance of a capacitor falls with increasing frequency, so that at high frequencies most of the voltage signal will be dropped across the internal resistance R.

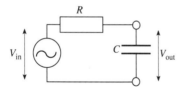

Figure 142

If we write Z for the impedance of the capacitor, the ratio v_{out}/v_{in} is given by $Z/(Z + R)$. Taking $Z = 1/i\omega C$, we have

$$\frac{v_{out}}{v_{in}} = \frac{1/i\omega C}{R + 1/i\omega C} = \frac{1}{1 + i\omega RC}.$$

The factor by which the power is reduced is thus

$$|v_{out}/v_{in}|^2 = \frac{1}{1 + \omega^2 R^2 C^2}.$$

For an attenuation of 3 dB at 1 Hz we require that this factor should be equal to $10^{-0.3} = 0.5012$. Hence

$$\omega RC = (0.5012^{-1} - 1)^{1/2} = 0.9976.$$

▶ Taking $\omega = 2\pi\,s^{-1}$ and $R = 197\,k\Omega$ gives $C = 8.06 \times 10^{-7}\,F = 0.81\,\mu F$. At
▶ 50 Hz, $\omega = 100\pi\,s^{-1}$ so $(1 + \omega^2 R^2 C^2)^{-1} = 4.02 \times 10^{-4}$ which corresponds
▶ to an attenuation of $-10\log_{10}(4.02 \times 10^{-4})\,dB = 34\,dB$.

[The choice of the '3 dB point' is based on the fact that $10^{-0.3}$ is very close to 1/2. If it were exactly equal to 1/2, the 3 dB point would occur at the frequency at which $\omega RC = 1$. The error in calculating the appropriate value of C using this approximation would be only 0.2%. We can also approximate the second part of the problem by noting that at a frequency 50 times greater, ωRC will be close to 50, and since $50^2 \gg 1$ we can neglect the 1 in the denominator and write the attenuation as approximately $1/50^2 = 4 \times 10^{-4}$.]

Problem 154

Determine the voltage across and the charge stored in each of the capacitors shown in figure 143.

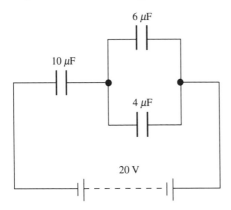

Figure 143

Solution

The 6 μF and 4 μF capacitors connected in parallel have a combined capacitance of $6 + 4 = 10\,\mu F$. The total capacitance of two 10 μF capacitors connected in series is $(1/10 + 1/10)^{-1} = 5\,\mu F$. Thus the total charge withdrawn from the positive terminal of the battery must be
▶ $5 \times 20 = 100\,\mu C$. This must be equal to the charge on the 10 μF capacitor,
▶ so that voltage across this capacitor is $100/10 = 10\,V$. The remaining 10 V must be dropped across the 6 μF and 4 μF capacitors, so the charge on the
▶ 6 μF capacitor is $6 \times 10 = 60\,\mu C$ and the charge on the 4 μF capacitor is
▶ $4 \times 10 = 40\,\mu C$. To summarise, we have the results shown in Table 6.

Table 6

	Voltage	Charge
10 μF capacitor	10 V	100 μC
6 μF capacitor	10 V	60 μC
4 μF capacitor	10 V	40 μC

Problem 155

A bridge for measuring the inductance L_1 and resistance R_1 of a coil is shown in figure 144. R_2 and R_3 are variable resistors, and C_3 and C_4 are fixed capacitors. Show that the conditions for the bridge to be balanced (no signal at the detector) are independent of the frequency of the alternating voltage source V.

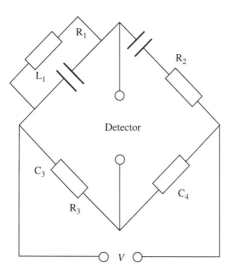

Figure 144

The bridge is found to be balanced with the following values:

$C_3 = 8\ \mu$F,
$C_4 = 5\ \mu$F,
$R_2 = 0.5\ \Omega$,
$R_3 = 3.2\ k\Omega$.

What are the values of L_1 and R_1?

Solution

When the circuit is balanced, the voltages at the two terminals of the detector are equal. If we write Z_1 for the total complex impedance of the inductor L_1 and the resistor R_1, and define Z_2, Z_3 and Z_4 similarly, the condition for balance is

$$\frac{Z_1}{Z_1 + Z_2} = \frac{Z_3}{Z_3 + Z_4},$$

which can be rewritten as

$$\frac{Z_1}{Z_2} = \frac{Z_3}{Z_4}$$

or

$$Z_1 Z_4 = Z_2 Z_3.$$

Now we have

$$Z_1 = R_1 + i\omega L_1, \qquad Z_2 = R_2,$$
$$Z_3 = R_3 + \frac{1}{i\omega C_3}, \qquad Z_4 = \frac{1}{i\omega C_4},$$

so

$$\frac{R_1}{i\omega C_4} + \frac{L_1}{C_4} = R_2 R_3 + \frac{R_2}{i\omega C_3}.$$

Equating real and imaginary parts of this expression gives

$$R_1 = \frac{R_2 C_4}{C_3} \text{ and } L_1 = C_4 R_3 R_2.$$

These conditions for balance are independent of the frequency ω as required. [This kind of bridge circuit is called an *Owen bridge*.]
▶   Substituting the values given for C_3, C_4, R_2 and R_3 gives $R_1 = 0.31\,\Omega$
▶   and $L_1 = 8.0\,\text{mH}$.

Problem 156

A lossy capacitor C_1 behaves as though it had a resistance R_1 in parallel with it. Using the bridge circuit shown in figure 145, a balance is obtained at 150 Hz when $R_3 = R_4 = 500\,\Omega$, $C_2 = 1\,\mu\text{F}$, $R_2 = 300\,\Omega$. Find the values of C_1 and R_1.

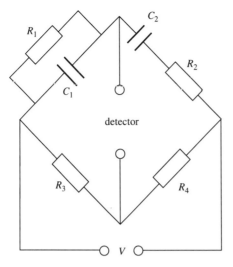

Figure 145

Solution

When the bridge is balanced, the complex impedances in the arms of the bridge satisfy the relationship

$$\frac{Z_2}{Z_1} = \frac{Z_4}{Z_3}.$$

Now we have

$$Z_2 = R_2 + \frac{1}{i\omega C_2} \quad \text{and} \quad \frac{1}{Z_1} = \frac{1}{R_1} + i\omega C_1,$$

so the condition for balance is

$$\frac{R_4}{R_3} = \frac{R_2}{R_1} + \frac{C_1}{C_2} + i\left(\omega C_1 R_2 - \frac{1}{\omega C_2 R_1}\right).$$

Equating real and imaginary parts gives

$$\frac{R_4}{R_3} = \frac{R_2}{R_1} + \frac{C_1}{C_2} \tag{1}$$

and

$$\omega^2 C_1 C_2 R_1 R_2 = 1. \tag{2}$$

Rearranging equation (2) to obtain an expression for $1/R_1$ gives

$$\frac{1}{R_1} = \omega^2 C_1 C_2 R_2,$$ (3)

and substituting this into equation (1) gives

$$\frac{C_1}{C_2} = \frac{R_4}{R_3} - \omega^2 C_1 C_2 R_2^2.$$ (4)

Equation (4) can now be rearranged to give an expression for C_1,

$$C_1 = \frac{R_4/R_3}{\omega^2 C_2 R_2^2 + 1/C_2},$$

▶ and substitution of the values given in the problem yields $C_1 = 0.93 \,\mu\text{F}$.
▶ Substitution of this value into equation (3) gives $R_1 = 4.1 \,\text{k}\Omega$.
[The calculation can be simplified by noting that, since $R_3 = R_4$, Z_1 must equal Z_2. Using $Z_2 = R_2 + 1/i\omega C_2$ and taking $R_2 = 300 \,\Omega$, $\omega = 300 \,\pi\,\text{s}^{-1}$ and $C_2 = 10^{-6}\,\text{F}$ gives

$$Z_2 = (300 - 1061i) \,\Omega.$$

Since $1/Z_1 = 1/R_1 + i\omega C_1$, we need to calculate the reciprocal of Z_2. It is

$$1/Z_2 = (2.468 + 8.727i) \times 10^{-4}\,\text{S},$$

so $R_1 = 1/(2.468 \times 10^{-4}\,\text{S}) = 4.1 \,\text{k}\Omega$ and $C_1 = (8.727 \times 10^{-4})/(300\,\pi)\text{F}$ $= 0.93 \,\mu\text{F}$ as before.]

Problem 157

A voltage $V_{AB} = V_0 \cos \omega t$, where V_0 is a real amplitude, is applied between the points A and B in the network shown in figure 146. Given that

$$C = \frac{1}{\omega R \sqrt{3}}$$

and

$$L = \frac{R\sqrt{3}}{\omega},$$

(i) calculate the total impedance between A and B,
(ii) verify that voltages of equal amplitude are developed between the point X and the points A, Y and Z,
(iii) determine the phases of these three voltages relative to V_{AB}.

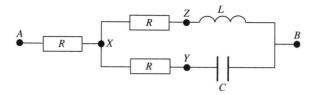

Figure 146

Solution

(i) The complex impedance Z of an inductor L is $i\omega L$, so if $L = R\sqrt{3}/\omega$, $Z = iR\sqrt{3}$.

The complex impedance Z of a capacitor C is $1/i\omega C$, so if $C = (\omega R\sqrt{3})^{-1}$, $Z = R\sqrt{3}/i = -iR\sqrt{3}$. Thus the complex impedance Z_U of the upper branch of the network between X and B is $R(1 + i\sqrt{3})$, and the complex impedance Z_L of the lower branch is $R(1 - i\sqrt{3})$. The complex impedance of Z_U in parallel with Z_L is

$$\frac{Z_U Z_L}{Z_U + Z_L} = \frac{R^2(1 + i\sqrt{3})(1 - i\sqrt{3})}{2R} = \frac{4R^2}{2R} = 2R.$$

▶   Thus the total impedance between A and B is $3R$.

(ii and iii) It is easier to solve these together, finding the amplitudes and phases of the voltages V_{AX}, V_{YX} and V_{ZX} simultaneously. We will use the complex exponential notation and, for convenience, set the voltage at B to zero. The voltage at A can thus be written as $V_0(0)$, by which we mean a voltage of amplitude V_0 and phase angle zero, and we will use the notation V_{AB} to refer to $V_A - V_B$, i.e. the voltage at A minus the voltage at B, etc.

Since the impedance between B and A is $3R$ and the impedance between B and X is $2R$ (both real), the voltage at X is given quite simply by

$$V_X = \frac{2}{3}V_A = \frac{2}{3}V_0(0).$$

Considering the upper branch between X and B, we have a potential divider so that

$$\frac{V_Z}{V_X} = \frac{iR\sqrt{3}}{R(1 + i\sqrt{3})} = \frac{i\sqrt{3}}{1 + i\sqrt{3}}.$$

The numerator of this expression has an amplitude of $\sqrt{3}$ and a phase angle of $90°$, and the denominator has an amplitude of $(1^2 + [\sqrt{3}]^2)^{1/2} = 2$

and a phase angle of $\tan^{-1}(\sqrt{3}/1) = 60°$. The whole expression therefore has an amplitude of $\sqrt{3}/2$ and a phase angle of $90° - 60° = 30°$. Thus

$$\frac{V_Z}{V_X} = \frac{\sqrt{3}}{2}(30°),$$

so

$$V_Z = \frac{V_0}{\sqrt{3}}(30°).$$

Similarly, we can write

$$\frac{V_Y}{V_X} = \frac{-iR\sqrt{3}}{R(1 - i\sqrt{3})} = \frac{-i\sqrt{3}}{1 - i\sqrt{3}} = \frac{\sqrt{3}}{2}(-30°),$$

so

$$V_Y = \frac{V_0}{\sqrt{3}}(-30°).$$

We can now calculate the voltage differences:

▶ $V_{AX} = V_A - V_X = V_0 - 2V_0/3 = V_0/3$, which, since it has a phase angle of zero, is in phase with the voltage V_{AB}.

$$V_{YX} = V_Y - V_X = \frac{V_0}{\sqrt{3}}(-30°) - \frac{2V_0}{3}.$$

This can be visualised on a phasor diagram, as shown in figure 147.

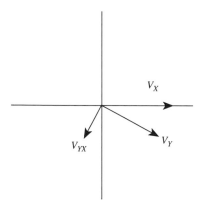

Figure 147

The real part of this expression is

$$\frac{V_0}{\sqrt{3}}\cos(-30°) - \frac{2V_0}{3} = -\frac{V_0}{6},$$

and the imaginary part is

$$\frac{V_0}{\sqrt{3}} \sin{(-30°)} = \frac{-V_0}{2\sqrt{3}}.$$

Thus the amplitude is

$$V_0 \left(\left[\frac{1}{6} \right]^2 + \left[-\frac{1}{2\sqrt{3}} \right]^2 \right)^{1/2} = \frac{V_0}{3}$$

and the phase is

$$\arctan \frac{-\dfrac{1}{2\sqrt{3}}}{-\dfrac{1}{6}} = -120°.$$

Thus

▶ $$V_{YX} = \frac{V_0}{3}(-120°).$$

Finally,

$$V_{ZX} = V_Z - V_X = \frac{V_0}{\sqrt{3}}(30°) - \frac{2V_0}{3}.$$

Again, we can show this on a phasor diagram (figure 148).

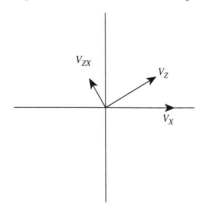

Figure 148

The real part of this is

$$\frac{V_0}{\sqrt{3}} \cos{(30°)} - \frac{2V_0}{3} = -\frac{V_0}{6}$$

and the imaginary part is

$$\frac{V_0}{\sqrt{3}} \sin (30°) = \frac{V_0}{2\sqrt{3}}.$$

The amplitude of V_{ZX} is again $V_0/3$, and the phase is $+120°$, so

▶ $$V_{ZX} = \frac{V_0}{3}(120°).$$

Problem 158

An electrical circuit consists of a resistance R, inductance L and capacitance C in series. If a charge is put on the capacitor at some instant, determine the condition that V_C, the voltage across the capacitor, is subsequently oscillatory. Assuming that this condition is satisfied, derive an expression for the time T for the amplitude of V_C to drop by a factor of e.

An external voltage source of variable frequency is introduced into the circuit in series with the other components. Show that, for small R, the width of the resonance (defined as the angular frequency range for which the amplitude of V_C is greater than $1/\sqrt{2}$ of its resonance value) is approximately $2/T$.

Solution

Figure 149 shows the circuit with no external voltage source.

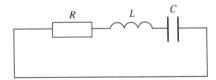

Figure 149

The e.m.f. across a resistor of resistance R carrying a current I is RI, and the e.m.f. across an inductor of inductance L carrying a current I is $L\, dI/dt$. A capacitor of capacitance C holding a charge Q has an e.m.f. Q/C across it, and since $I = dQ/dt$ the e.m.f. across the capacitor is given by

$$\frac{1}{C}\int I\, dt.$$

Since the three components are connected in series, the same current I must flow through each, and the sum of the e.m.f.s across all the components must be zero, so we must have

$$RI + L\frac{dI}{dt} + \frac{1}{C}\int I \, dt = 0.$$

Differentiating this expression with respect to time gives a differential equation for the current I:

$$L\frac{d^2I}{dt^2} + R\frac{dI}{dt} + \frac{1}{C}I = 0.$$

Now if the e.m.f. across the capacitor is to vary in an oscillatory manner, the current I through the circuit will also have an oscillatory behaviour. We will therefore look for an oscillatory solution of the form

$$I = I_0 \exp(-i\omega t),$$

in which case

$$dI/dt = -i\omega I$$

and

$$d^2I/dt^2 = -\omega^2 I.$$

Substituting into the differential equation gives

$$-L\omega^2 I - i\omega RI + \frac{I}{C} = 0,$$

which is a quadratic equation in ω. The solutions of this equation are

$$\omega = \frac{-iR \pm \sqrt{(-R^2 + 4L/C)}}{2L},$$

so the condition that the current (and hence V_C) is oscillatory, which is equivalent to saying that ω must have a real part, is that

▶ $$R^2 < \frac{4L}{C}.$$

If this condition is satisfied, we may write $\omega = a - iR/2L$ where a is real. Thus the current I has the form

$$I = I_0 \exp(-i\omega t) = I_0 \exp(-iat)\exp(-Rt/2L).$$

The amplitude of the oscillatory current, and therefore of the e.m.f. V_C, varies with time as $\exp(-Rt/2L)$, and thus the time T for this amplitude
▶ to fall by a factor of e is given by $T = 2L/R$.

Now we insert an a.c. voltage source into the circuit, as shown in figure 150.

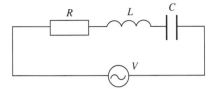

Figure 150

Since this is an a.c. problem, we can use complex impedances. The total impedance of the circuit $Z_T = Z_R + Z_L + Z_C$, where Z_L is the complex impedance of the inductor, and so on. The current in the circuit is given by $I = V/Z_T$, and the e.m.f. across the capacitor is given by $V_C = IZ_C$. Thus we have

$$\frac{V_C}{V} = \frac{Z_C}{Z_R + Z_L + Z_C} = \frac{\dfrac{1}{i\omega C}}{R + i\omega L + \dfrac{1}{i\omega C}} = \frac{1}{1 - \omega^2 CL + iR\omega C}.$$

Since R is small, $|V_C/V|$ will attain its maximum value when $1 - \omega^2 CL \approx 0$, i.e. at an angular frequency $\omega_0 = 1\sqrt{(LC)}$. The maximum value of $|V_C/V|$ is clearly $1/(R\omega_0 C)$.

Also since R is small, the resonance will be narrow so that the frequency ω will not differ very much from ω_0. We can thus neglect changes in the imaginary part of the denominator of our expression for V_C/V. Since we are looking for the frequencies ω at which $|V/V_C|$ is $1/\sqrt{2}$ times its maximum value, these must occur when the real part of the denominator is equal to $\pm$ the imaginary part. Thus

$$(1 - \omega^2 CL) = \pm R\omega_0 C,$$

which can be rearranged, using the fact that $\omega_0 = 1/\sqrt{(CL)}$, as

$$\left(\frac{\omega}{\omega_0}\right)^2 = 1 \pm R\omega_0 C.$$

If $R\omega_0 C \ll 1$ we can use the binomial expansion to take the square root of this expression:

$$\left(\frac{\omega}{\omega_0}\right) \approx 1 \pm \frac{R\omega_0 C}{2}.$$

The two solutions for ω are thus $\omega_0 + R\omega_0^2 C/2$ and $\omega_0 - R\omega_0^2 C/2$, so that the width of the resonance is $\Delta\omega = R\omega_0^2 C = R/L = 2/T$, as required.

Problem 159

A capacitor C, which may be considered to be loss-free, is connected with an inductor to form a closed loop. The inductor behaves as a pure inductance L in series with a resistor R. If, when the circuit is first connected the capacitor is charged, derive the condition for oscillatory decay.

Derive an expression for the *quality factor* Q_1 of the circuit by considering the decay of the oscillation, using the result that the amplitude falls by a factor of e in Q_1/π periods.

A quality factor Q_2 can also be defined in terms of the circuit parameters as $\omega_0 L/R$, where $\omega_0^2 = 1/LC$. If $C = 2\ \mu F$ and $L = 1\ mH$, calculate the maximum value of R such that Q_1 and Q_2 agree to better than 1 part in 10^3.

Solution

The first part of the solution proceeds exactly as in problem 158, and leads to the result that the condition for oscillation is

▶
$$R^2 < \frac{4L}{C}.$$

If this condition is satisfied, the current I flowing in the circuit can be described by

$$I = I_0 \exp(-i\omega t),$$

where ω is complex. As we saw in problem 158, the real part of ω is

$$\frac{\sqrt{(4L/C - R^2)}}{2L}$$

and the imaginary part is

$$-\frac{R}{2L}.$$

If we write this as $\omega = a - ib$, the decay has the form

$$I = I_0 \exp(-iat)\exp(-bt),$$

so the period is $2\pi/a$ and the time taken for the amplitude to fall by a factor of e is $1/b$. Thus the number of cycles required for the amplitude to fall by a factor of e is $a/2\pi b$, so we have

$$Q_1 = \pi\frac{a}{2\pi b} = \frac{a}{2b}.$$

Substituting our expressions for the real and imaginary parts a and b of ω thus gives

$$\blacktriangleright \qquad Q_1 = \frac{\sqrt{(4L/C - R^2)}}{2R}.$$

The second expression for Q can be written as

$$Q_2 = \frac{\omega_0 L}{R} = \frac{1}{R}\sqrt{\frac{L}{C}},$$

so the ratio of the two expressions is

$$\frac{Q_1}{Q_2} = \frac{\sqrt{(4L/C - R^2)}}{2}\sqrt{\frac{C}{L}} = \sqrt{\left(1 - \frac{R^2 C}{4L}\right)}.$$

Since we require this ratio to be equal to 1 to within 1 part in 10^3, we are justified in using a binomial expansion to approximate the square root:

$$\frac{Q_1}{Q_2} \approx 1 - \frac{R^2 C}{8L}.$$

Thus if the two expressions are to agree to better than 1 part in 10^3, we must have

$$\frac{R^2 C}{8L} < 10^{-3}.$$

$\blacktriangleright$ Substituting the values given in the problem yields $R < 2\,\Omega$.

Problem 160

A series LCR circuit has resonant frequency ω_0 and a large quality factor Q. Write down in terms of R, ω, ω_0 and Q (a) its impedance at resonance, (b) its impedance at the half-power points, and (c) the approximate forms of its impedance at low and high frequencies.

Solution

(a) The complex impedance Z of a resistance R, inductance L and capacitance C in series is

$$Z = R + i\omega L + \frac{1}{i\omega C} = R + i\omega L - \frac{i}{\omega C}.$$

$\blacktriangleright$ At resonance the imaginary part of Z is zero, so that $Z = R$.

(b) At the half-power points $|Z|$ is a factor $\sqrt{2}$ greater than its value at resonance. Since the real part of Z has a fixed value of R, this means that the imaginary part at the half-power points must be $\pm R$, so the values of ▶ Z are $R(1 \pm i)$.

(c) We know (e.g. from problem 159) that

$$\omega_0 = \frac{1}{\sqrt{(LC)}} \quad \text{and} \quad Q = \frac{1}{R}\sqrt{\frac{L}{C}},$$

so $C = 1/Q\omega_0 R$ and $L = QR/\omega_0$. At sufficiently low frequencies, Z is dominated by the effect of the capacitor, so $Z \approx -i/\omega C$. Thus at low frequencies

▶ $$Z \approx \frac{-iQR\omega_0}{\omega}.$$

At sufficiently high frequencies, the impedance is dominated by the inductor, so $Z \approx i\omega L$ which can be written as

▶ $$Z \approx \frac{iQR\omega}{\omega_0}.$$

Problem 161

In the circuit shown in figure 151 the inductor has a self-inductance L and the three resistors have the same resistance R. The switch is closed at $t = 0$. Obtain expressions for the currents I_1 and I_2 as functions of time, and illustrate these variations on a labelled graph.

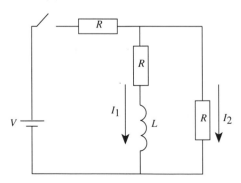

Figure 151

Solution

It will simplify the problem if we assign a fixed value to the voltage at some point in the circuit, so that we can deal with voltages rather than voltage differences. Let us set the negative terminal of the battery at zero, so that the positive terminal is at V. It will also simplify the problem if we introduce variables I_0 for the total current drawn from the battery, and V' for the voltage at the point where the three resistors are connected together, as shown in figure 152.

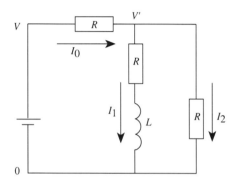

Figure 152

For the resistor in which current I_0 is flowing, we have

$$I_0 = \frac{V - V'}{R}. \tag{1}$$

For the resistor in which current I_2 is flowing, we have

$$I_2 = \frac{V'}{R}. \tag{2}$$

For the third resistor, we have

$$V' = I_1 R + L\frac{dI_1}{dt}. \tag{3}$$

We can eliminate I_0 since it must be equal to $I_1 + I_2$. Applying this to (1) and rearranging the equation gives

$$V' = V - R(I_1 + I_2). \tag{4}$$

Now we can eliminate V' by combining (2) and (4):

$$V = 2RI_2 + RI_1. \tag{5}$$

Similarly, combining (3) and (4) gives

$$V = 2RI_1 + RI_2 + L\frac{dI_1}{dt}. \tag{6}$$

The simplest way of proceeding from here is to combine (5) and (6) to obtain a differential equation containing either I_1 or I_2, but not both. It will be easier to work with I_1, which gives

$$\frac{3RI_1}{2} + L\frac{dI_1}{dt} = \frac{V}{2}. \tag{7}$$

The steady state solution to this equation is found by putting $dI_1/dt = 0$, which gives $I_1 = V/3R$ (this result is obvious from inspection of the circuit diagram, since at steady state we may replace the inductor by a short circuit). We know that I_1 must be zero at time $t = 0$ (since the inductor will oppose the attempt to increase the current flowing through it), so we can try a solution of the form

$$I_1 = \frac{V}{3R}(1 - \exp[-\alpha t]),$$

which clearly has the right behaviour at $t = 0$ and $t = \infty$. Using this expression for I_1 gives

$$\frac{dI_1}{dt} = \frac{V\alpha}{3R}\exp(-\alpha t),$$

and substituting these expressions into (7) gives

$$\frac{V}{2}(1 - \exp[-\alpha t]) + \frac{LV\alpha}{3R}\exp(-\alpha t) = \frac{V}{2},$$

which is satisfied if $\alpha = 3R/2L$. Thus the solution for I_1 is

$$\blacktriangleright \qquad I_1 = \frac{V}{3R}\left(1 - \exp\left[-\frac{3Rt}{2L}\right]\right),$$

and substitution into (5) gives the solution for I_2 as

$$\blacktriangleright \qquad I_2 = \frac{V}{3R}\left(1 + \frac{1}{2}\exp\left[-\frac{3Rt}{2L}\right]\right).$$

The value of I_2 at $t = 0$ is thus $V/2R$, and its steady state value at $t = \infty$ is $V/3R$. The graphs of I_1 and I_2 as functions of time are shown in figure 153.

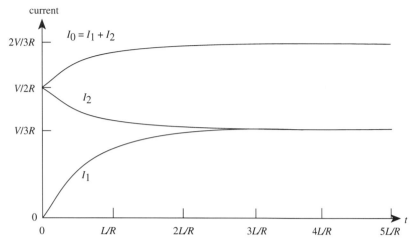

Figure 153

Problem 162

A solenoid with self-inductance $L = 2$ H and carrying a steady current of 1 A has the current source suddenly disconnected. What is the minimum capacitance which should be connected across the terminals of the solenoid in order to prevent the potential difference generated by the collapse of the magnetic field from rising above 300 V?

Solution

The simplest approach is to recall that the energy stored by an inductor carrying a current I is $LI^2/2$ and that the energy stored by a capacitor charged to a potential V is $CV^2/2$. Initially (before the current source is disconnected) $I = I_0 = 1$ A and $V = 0$. The maximum potential difference V_{max} is thus given by

$$\frac{1}{2}CV_{max}^2 = \frac{1}{2}LI_0^2,$$

which can be rearranged to give

$$C = L\left(\frac{I_0}{V_{max}}\right)^2.$$

▶   Putting $L = 2$ H, $I_0 = 1$ A and $V_{max} = 300$ V gives $C = 22\ \mu$F.

[Another approach to the problem is to recognise that, once the current source has been disconnected, the circuit is an LC resonant circuit in which the current will oscillate at angular frequency $\omega = (LC)^{-1/2}$. Since the current must be equal to I_0 at $t = 0$, we may write

$$I = I_0 \cos \frac{t}{\sqrt{(LC)}}.$$

Now the e.m.f. V across an inductor in which the current is changing is given by $V = -L\,dI/dt$, so we have

$$V = \frac{LI_0}{\sqrt{(LC)}} \sin \frac{t}{\sqrt{(LC)}} = I_0 \sqrt{\frac{L}{C}} \sin \frac{t}{\sqrt{(LC)}}.$$

Thus we see that the maximum e.m.f. $V_{\max}$ and the maximum current I_0 are related by

$$\frac{1}{2}CV_{\max}^2 = \frac{1}{2}LI_0^2$$

as before.]

Problem 163

Figure 154 shows an amplifier constructed from a field-effect transistor with mutual conductance $g_m = 2 \times 10^{-3}$ S.
    For small a.c. signals v_{in} calculate:
    (i) the voltage gain of the amplifier;
    (ii) the input impedance.
(The capacitors C_1 and C_2 can be assumed to have negligible a.c. impedances.)

Solution

To a good approximation, we can assume that the gate of the FET draws no current and that there is infinite impedance between the drain and the source. The mutual conductance g_m is defined by

$$i_{ds} = g_m(v_g - v_s),$$

where i_{ds} is the (a.c.) current flowing in to the source and out of the drain, and v_g and v_s are the (a.c.) voltages at the gate and drain respectively. We can thus draw the equivalent circuit for small a.c. signals, as shown in figure 155.
    As a first approximation, let us ignore the current flowing through the

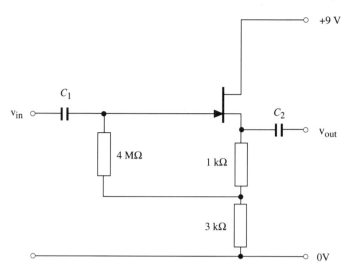

Figure 154

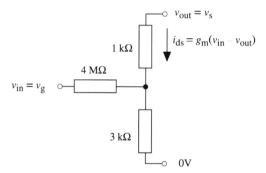

Figure 155

$4\,\text{M}\Omega$ resistor. The current i_{ds} flows through a total resistance of $4\,\text{k}\Omega$ to give a voltage drop of v_{out}, so numerically we can put

$$4 \times 10^3 \times 2 \times 10^{-3}\,(v_{in} - v_{out}) = v_{out},$$

which can be simplified to

$$8\,v_{in} = 9\,v_{out},$$

▶ so that the voltage gain v_{out}/v_{in} is $8/9 = 0.889$.

Using this same approximation, the voltage at the point where the three resistors meet is $3v_{out}/4$ (the $1\,\text{k}\Omega$ and $3\,\text{k}\Omega$ resistors are just acting as a potential divider) which can be written as $2v_{in}/3$, so the voltage across the $4\,\text{M}\Omega$ resistor is $v_{in}/3$. The current flowing in this resistor is

▶ numerically $v_{in}/(12 \times 10^6)$, so the input impedance is $12\,\text{M}\Omega$.

[Let us now calculate the results exactly. It is probably easiest to begin with our approximate results, and put

$$v_{out} = \frac{8}{9}v_{in}(1 + \alpha) \text{ and } v_? = \frac{2}{3}v_{in}(1 + \beta),$$

where v_2 is the voltage at the point where the three resistors meet, and we expect $|\alpha|$ and $|\beta|$ to be much less than 1.

The current i_{ds} is given by $g_m(v_{in} - v_{out})$ which can be written numerically as

$$2 \times 10^{-3}v_{in}(1 - 8[1 + \alpha]/9) = 2 \times 10^{-3}v_{in}(1/9 - 8\alpha/9).$$

The voltage dropped in the 1 kΩ resistor is numerically a thousand times this value, so we can write

$$\frac{8}{9}v_{in}(1 + \alpha) - \frac{2}{3}v_{in}(1 + \beta) = 2v_{in}\left(\frac{1}{9} - \frac{8\alpha}{9}\right),$$

which simplifies to $\beta = 4\alpha$.

The current flowing in the 4 MΩ resistor is numerically equal to

$$\frac{v_{in} - \frac{2}{3}v_{in}(1 + \beta)}{4 \times 10^6} = 2.5 \times 10^{-7}v_{in}\left(\frac{1}{3} - \frac{8\alpha}{3}\right),$$

so the total current flowing in the 3 kΩ resistor is numerically equal to

$$2 \times 10^{-3}v_{in}\left(\frac{1}{9} - \frac{8\alpha}{9}\right) + 2.5 \times 10^{-7}v_{in}\left(\frac{1}{3} - \frac{8\alpha}{3}\right),$$

which must be equal to $v_2/(3\text{ k}\Omega)$. Expressing this numerically, and multiplying through by 3×10^3,

$$\frac{2}{3}v_{in}(1 + 4\alpha) = 6v_{in}\left(\frac{1}{9} - \frac{8\alpha}{9}\right) + 7.5 \times 10^{-4}v_{in}\left(\frac{1}{3} - \frac{8\alpha}{3}\right).$$

This can easily be solved for α to give 3.12×10^{-5}. The gain is thus $8(1 + 3.12 \times 10^{-5})/9$, which is negligibly different from the value we calculated before. Substituting this value of α into our expression for the current flowing in the 4 MΩ resistor gives $8.33125 \times 10^{-8}v_{in}$, so the input impedance is 12.003 MΩ, which is also negligibly different from the previous value. Our initial assumption that the input current could be neglected in calculating the various voltages in the circuit was thus entirely justified.]

Problem 164

Figure 156 shows a typical low-power npn bipolar transistor of current gain $\beta = \partial I_c / \partial I_b$ and mutual conductance $g_m = \partial I_c / \partial V_{be}$ connected for use as an emitter follower. The transistor is fed by an a.c. signal from a source of small e.m.f. v_s and negligible impedance, and drives a load of impedance R_L. Biassing components are not shown and the impedances of the capacitors can be assumed to be negligible.

(i) Derive expressions for the input impedance of the circuit as seen by the source and the output impedance as seen by the load.

(ii) Derive an expression for the a.c. voltage gain of the circuit and comment on its numerical value.

(iii) Suggest a suitable biassing arrangement and give rough values of the circuit components appropriate for a supply voltage $V_0 = 9$ V and a quiescent emitter current of 1 mA.

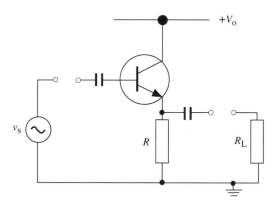

Figure 156

Solution

(i and ii) Since we are dealing with a.c. signals, we can put

$$i_c = \beta i_b = g_m(v_b - v_e).$$

The emitter current i_e is $i_b + i_c = i_b(1 + \beta)$. We can show these currents and voltages on a modified circuit diagram (figure 157) from which the capacitors have been removed (since they have negligible a.c. impedance).

It is clear that the emitter current i_e flows through R and R_L in parallel,

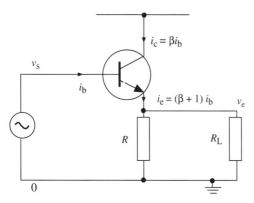

Figure 157

so if we write r for their combined resistance (i.e. $1/r = 1/R + 1/R_L$), we must have

$$v_e = i_e r = (\beta + 1)i_b r.$$

Combining this with the fact that $\beta i_b = g_m(v_b - v_e)$, and noting that $v_b = v_s$, gives

$$\beta i_b = g_m(v_s - [\beta + 1]i_b r),$$

which can be rearranged as

$$i_b(\beta + [\beta + 1]g_m r) = g_m v_s,$$

so the input impedance $R_{in} = v_s/i_b$ is

$$R_{in} = \frac{\beta + (\beta + 1)g_m r}{g_m}.$$

[In practice, $g_m r \gg 1$ and $\beta \gg 1$, so this can be written approximately as $R_{in} \approx \beta r$. This will be large, which is what is needed in a follower circuit.]

It is perhaps marginally easier to find the output impedance by first finding the a.c. voltage gain. This is defined as

$$G = \frac{v_e}{v_s}.$$

If we put $v_e = (\beta + 1)i_b r$, this becomes

$$G = \frac{(\beta + 1)i_b r}{v_s},$$

and we can substitute the reciprocal of our expression for R_{in} in place of i_b/v_s to give

▶ $$G = \frac{(\beta + 1)g_m r}{\beta + (\beta + 1)g_m r}.$$

Since $g_m r \gg 1$ and $\beta \gg 1$, this will be very close to (just less than) unity, which is another of the requirements of a follower circuit.

To find the output impedance R_{out} of the circuit, it is helpful to consider a model of the amplifier's behaviour, as shown in figure 158.

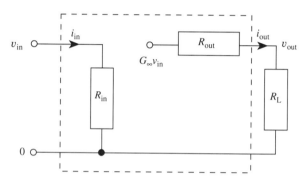

Figure 158

G_∞ is the gain of the amplifier when there is no load resistor (i.e. when R_L is infinite). The output current i_{out} is given by

$$i_{out} = \frac{G_\infty v_{in}}{R_{out} + R_L}$$

and it can also be written as v_{out}/R_L. Equating these two expressions, and writing G for the amplifier gain v_{out}/v_{in} when the load resistor is connected, we obtain

$$\frac{G_\infty}{R_{out} + R_L} = \frac{G}{R_L},$$

which can be rearranged to give

$$R_{out} = \left(\frac{G_\infty}{G} - 1 \right) R_L.$$

We have already shown that the gain G is given by

$$G = \frac{(\beta + 1)g_m r}{\beta + (\beta + 1)g_m r},$$

so

$$\frac{1}{G} = 1 + \frac{\beta}{(\beta + 1)g_m r}.$$

Since $g_m r \gg 1$ and $(\beta + 1) \approx \beta$, we can use a binomial expansion to derive

$$G \approx 1 - \frac{1}{g_m r}.$$

G_∞ is the value of G when R_L is infinite, i.e. when $r = R$, so using the same approximations,

$$G_\infty \approx 1 - \frac{1}{g_m R}.$$

Hence

$$R_{out} \approx R_L \left(\frac{1 - \frac{1}{g_m R}}{1 - \frac{1}{g_m r}} - 1 \right) \approx R_L \left(\frac{1}{g_m r} - \frac{1}{g_m R} \right) = \frac{1}{g_m}.$$

▶ Thus the output impedance will be approximately $1/g_m$, which will be large, as required.

[If we do the calculation without approximations, we have

$$\frac{R_{out}}{R_L} = \frac{R(\beta + [\beta + 1]g_m r)}{r(\beta + [\beta + 1]g_m R)} - 1.$$

Using the fact that $R/r = 1 + R/R_L$ we can rewrite this as

$$\frac{R_{out}}{R_L} = \frac{\beta + \beta R/R_L + (\beta + 1)g_m R}{\beta + (\beta + 1)g_m R} - 1 = \frac{\beta R/R_L}{\beta + (\beta + 1)g_m R}.$$

The output impedance is thus

▶ $$R_{out} = \frac{\beta R}{\beta + (\beta + 1)g_m R}.]$$

(iii) A suitable biassing arrangement might be as shown in figure 159.

In order that the output should be able to swing as far in the positive direction as in the negative direction, the d.c. voltage V_e of the emitter should be close to 4.5 V. If the quiescent emitter current is 1 mA, this
▶ would require $R \approx 4.5$ kΩ. The base voltage V_b will be about 0.6 V higher than this, or about 5.1 V. The voltage dropped across R_1 is thus 3.9 V and the voltage dropped across R_2 is 5.1 V, so $R_1/R_2 = 3.9/5.1 = 0.76$. The resistance of R_1 in parallel with R_2 should be small compared

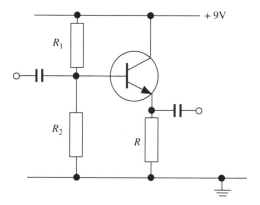

Figure 159

with the a.c. input impedance, which is of the order of βR, so if we assume that $\beta \approx 100$, it might be reasonable to choose R_1 and R_2 such that their parallel resistance is less than about 50 kΩ. Reasonable values might be $R_1 \approx 40$ kΩ and $R_2 \approx 50$ kΩ.

Problem 165

Figure 160 shows an amplifier using two identical npn bipolar transistors connected to form a long-tailed pair. The two inputs carry small-signal voltages v_1 and v_2. Each transistor has mutual conductance $g_m = \partial I_c / \partial V_{be}$.

(i) For the values of the resistors given find the value of the quiescent DC voltage at the collector of T_2.

(ii) Derive an expression for the small-signal voltage v_{out} in terms of v_1 and v_2.

(iii) Outline the advantages of replacing R_3 by a constant-current source.

(iv) Sketch a circuit suitable for use as a constant-current source in this context.

Solution

(i) Under quiescent conditions, the inputs are zero and the bases of the transistors can be assumed to be at 0 V. The voltage drop between the base and the emitter of a conducting transistor is typically 0.6 V, so the emitters will be at about -0.6 V, giving a voltage of 8.4 V across R_3. The current in R_3 is thus 1.79 mA, of which (by symmetry) half is contributed

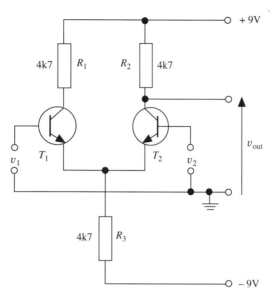

Figure 160

by T_2. Since the base current is very small compared with the emitter current, the collector current will be about 0.90 mA, so the voltage dropped across R_2 will be about 4.2 V. The voltage at the collector of T_2 will thus be about 4.8 V.

(ii) For the a.c. analysis, we can redraw the circuit as shown in figure 161.

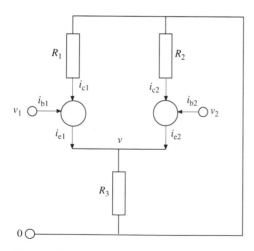

Figure 161

From the definition of g_m, we can write

$$i_{c1} = g_m(v_1 - v),$$
$$i_{c2} = g_m(v_2 - v),$$

where v is the a.c. voltage at the point where the emitters are connected together. The collector current i_c is given by βi_b and the emitter current i_e is $(\beta + 1)i_b$, so $i_e = (\beta + 1)i_c/\beta$. Thus

$$i_{e1} = \frac{g_m(\beta + 1)}{\beta}(v_1 - v) \text{ and } i_{e2} = \frac{g_m(\beta + 1)}{\beta}(v_2 - v).$$

The current in the 'tail' resistor R_3 is therefore

$$i_{e1} + i_{e2} = g'(v_1 - v + v_2 - v),$$

where $g' = g_m(\beta + 1)/\beta$. This current must also be equal to v/R_3, since the other end of R_3 is at zero a.c. voltage, so

$$g'(v_1 + v_2 - 2v) = \frac{v}{R}.$$

(Since $R_1 = R_2 = R_3$, we can just write R for their value.)

Rearranging this to obtain an expression for v in terms of v_1 and v_2 gives

$$v = \frac{g'R(v_1 + v_2)}{1 + 2g'R}.$$

We can now substitute this into our expression for i_{c2}:

$$i_{c2} = g_m(v_2 - v) = \frac{g_m(v_2 + v_2 g'R - v_1 g'R)}{1 + 2g'R}.$$

This current flows through the resistor R_2 from a.c. voltage zero to a.c. voltage v_{out} (at the collector of T_2), so $v_{out} = -Ri_{c2}$, giving

▶
$$v_{out} = \frac{g_m R}{1 + 2g'R}(v_1 g'R - v_2 g'R - v_2).$$

[We can calculate the numerical values in this expression reasonably accurately. The relationship between I_c and V_{be} in a transistor is given by the *Ebers–Moll equation*

$$I_c = I_s\left(\exp\left[\frac{eV_{be}}{kT}\right] - 1\right),$$

where I_s is the saturation current, e is the charge on the electron, k is the Boltzmann constant and T is the absolute temperature. Since $I_c \gg I_s$ for a transistor in the conduction region, the '−1' term can be ignored and the resulting exponential equation can be differentiated to give

$$\frac{\partial I_c}{\partial V_{be}} = \frac{eI_c}{kT}.$$

Thus $g_m = eI_c/kT = e\beta I_e/(\beta + 1)kT$. Taking $I_e = 0.90\,\text{mA}$, $T = 300\,\text{K}$ and $\beta = 100$ (the value is not critical as long as $\beta \gg 1$) gives $g_m = 0.034\,\text{S}$. g' thus equals $(\beta + 1)g_m/\beta = 0.035\,\text{S}$. With $R = 4700\,\Omega$, our expression for v_{out} becomes

$$v_{\text{out}} \approx 80(v_1 - v_2) - \frac{v_2}{2}.]$$

(iii) The circuit is intended to function as a differential amplifier, so that the output should be proportional to $(v_1 - v_2)$ with no dependence on the mean input signal $(v_1 + v_2)/2$. We can see that the circuit performs fairly well in this respect, since its differential-mode gain is about 80 and its common-mode gain is only about −0.5. However, it would be better if the common-mode gain could be reduced to zero, and this is the purpose of replacing R_3 by a constant-current source. We can see this as follows:

A constant (d.c.) current requires that the sum of the two (d.c.) emitter currents is constant, so that a small-signal current i_{e1} must be compensated by $i_{e2} = -i_{e1}$. The sum of the two collector currents must therefore be zero, which requires that $v = (v_1 + v_2)/2$. The current i_{c2} is therefore given by $g'(v_2 - v) = g'(v_2 - v_1)/2$, so the output voltage $v_{\text{out}} = -Ri_{c2} = Rg'(v_1 - v_2)/2$. In principle, then (i.e. assuming that the constant-current source is perfect), the common-mode gain has been reduced to zero.

(iv) A suitable constant-current source using the voltages available in the circuit could be as shown in figure 162.

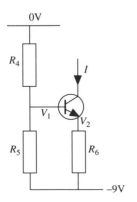

0V

R_4

I

V_1 V_2

R_5 R_6

−9V

Figure 162

If we require $I \approx 1.8$ mA and choose $R_6 =$ (say) 1 kΩ, V_2 must be -7.2 V. V_1 will be about 0.6 V greater than this, i.e. about -6.6 V, so the voltage dropped across R_4 is 6.6 V and that across R_5 is 2.4 V. The resistors R_4 and R_5 should therefore have a ratio of about 2.75.

Problem 166

In the generalised operational amplifier circuit shown in figure 163, Z_1, Z_2 and Z_3 are impedances. The operational amplifier has an input impedance greater than 10 MΩ and a gain equal to 10^5 up to 100 Hz but inversely proportional to frequency at higher frequencies.

(a) If $Z_1 = 1$ kΩ and $Z_2 = 100$ kΩ, draw a graph to show how the voltage ratio V_2/V_1 varies with frequency.

(b) If $Z_1 = 1$ kΩ and Z_2 consists of a parallel combination of a resistor $R_2 = 10$ kΩ and a capacitor $C = 1$ μF, draw a graph to show how V_2/V_1 varies with frequency.

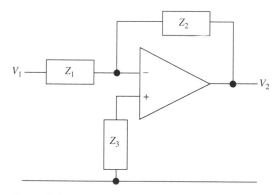

Figure 163

Solution

We will entirely neglect the input impedance of the operational amplifier, so that the inputs are assumed to draw no current. The positive terminal of the operational amplifier is thus at a voltage of zero.

If we write A for the gain of the operational amplifier, the voltage at its negative terminal must be $-V_2/A$. Equating the current in Z_1 to the current in Z_2 (since no current enters the amplifier) thus gives

$$\frac{V_1 + V_2/A}{Z_1} = \frac{-V_2/A - V_2}{Z_2}.$$

Problem 167

Give the truth table for the circuit shown in figure 166.

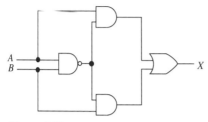

Figure 166

Solution

From left to right in figure 166, the logic gates are a NAND, two ANDS, and an OR.

To construct the truth table, it is helpful to label the intermediate connexions in the circuit, as shown in figure 167.

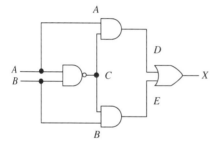

Figure 167

First we consider the output C of the NAND gate:

A	B	C
0	0	1
0	1	1
1	0	1
1	1	0

Next we consider the state D which is the output of an AND gate operating on A and C:

A	B	C	D
0	0	1	0
0	1	1	0
1	0	1	1
1	1	0	0

Similarly, *E* is the output of an AND gate operating on *B* and *C*:

A	B	C	D	E
0	0	1	0	0
0	1	1	0	1
1	0	1	1	0
1	1	0	0	0

▶ Finally, the output *X* is the result of an OR gate operating on *D* and *E*:

A	B	C	D	E	X
0	0	1	0	0	0
0	1	1	0	1	1
1	0	1	1	0	1
1	1	0	0	0	0

We recognise this as the exclusive-or (XOR) function.

[This result could also have been calculated using Boolean algebra, as follows:

$$C = \overline{A \cdot B};$$
$$D = A \cdot \overline{A \cdot B} = A \cdot (\bar{A} + \bar{B});$$
$$E = B \cdot \overline{A \cdot B} = B \cdot (\bar{A} + \bar{B});$$
$$X = D + E = (A + B) \cdot (\bar{A} + \bar{B})$$
$$= A \cdot \bar{A} + B \cdot \bar{B} + A \cdot \bar{B} + B \cdot \bar{A}$$
$$= A \cdot \bar{B} + B \cdot \bar{A},$$

which is the exclusive-or function.]

Problem 168

The following table gives the state of a J–K flip-flop after receipt of the *n*th clock pulse:

J	K	Q_n
0	0	Q_{n-1}
0	1	0
1	0	1
1	1	NOT Q_{n-1}

For the circuit shown in figure 168, deduce how the output *F* changes on receipt of a regular train of clock pulses, given that the initial state is $Q_1 = Q_2 = 0$.

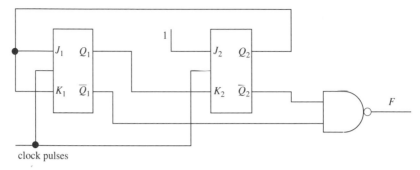

clock pulses

Figure 168

Solution

The output F is (NOT Q_1) NAND (NOT Q_2), which, by De Morgan's theorem, is Q_1 OR Q_2. Before the first clock pulse is received, we are told that $Q_1 = Q_2 = 0$. The output F is thus 0.

When the first clock pulse is received, the inputs J_1 and K_1 are both zero (since they are derived from Q_2), so the first flip-flop retains its state and Q_1 remains at 0. The inputs to the second flip-flop are $J_2 = 1$ (fixed) and $K_2 = 0$, so its output is set to $Q_2 = 1$. The output of the whole circuit is thus $F = 1$.

When the second clock pulse is received, J_1 and K_1 are both 1 so the flip-flop switches its state and Q_1 changes to 1. The inputs to the second flip-flop are $J_2 = 1$ and $K_2 = 0$ (the old value of Q_1), so its output is again set to $Q_2 = 1$, and the output of the whole circuit is $F = 1$.

When the third clock pulse is received, J_1 and K_1 are both 1 so the flip-flop again switches its state, reverting to $Q_1 = 0$. The inputs to the second flip-flop are $J_2 = 1$ and $K_2 = 1$ (the old value of Q_1), so this flip-flop also switches its state, giving $Q_2 = 0$. The output of the whole circuit is $F = 0$, and the circuit is now in exactly the same state as it was before the first clock pulse was received, so this behaviour will repeat itself after every three clock pulses. The output F will thus follow the pattern:

$$0 \quad 1 \quad 1 \quad 0 \quad 1 \quad 1 \quad \ldots$$

Thermodynamics

Problem 169

A well lagged wire of length L and cross-sectional area A has its ends maintained at temperatures T_1 and T_2. The thermal conductivity of the wire is given by

$$K = B + CT,$$

where T is the temperature and B and C are constants. What is the rate of flow of heat along the wire?

Solution

The rate at which heat is flowing through the wire is given by the definition of thermal conductivity as

$$\frac{dQ}{dt} = -KA\frac{dT}{dx}.$$

Since the wire is well lagged, we may assume that no heat enters or leaves it except at the ends, so dQ/dt must be constant. For convenience, let us put

$$D = \frac{1}{A}\frac{dQ}{dt},$$

where D is a constant. Substituting the expression given for K, we find

$$(B + CT)\frac{dT}{dx} = -D.$$

This differential equation can be solved by rearranging and integrating:

$$\int_{T_1}^{T_2}(B + CT)\,dT = -D\int_0^L dx.$$

This gives

$$B(T_2 - T_1) + \frac{C}{2}(T_2^2 - T_1^2) = -DL.$$

So

$$\frac{dQ}{dt} = AD = \frac{A}{L}\left(B[T_1 - T_2] + \frac{C}{2}[T_1^2 - T_2^2]\right).$$

This can conveniently be rearranged to give

$$\blacktriangleright \quad \frac{dQ}{dt} = \frac{A}{L}(T_1 - T_2)\left(B + \frac{C}{2}[T_1 + T_2]\right).$$

[We can see that this gives the right answer when $C = 0$ (i.e. when the conductivity does not change with temperature). We can also see that the answer is plausible when $C \neq 0$, since $B + C(T_1 + T_2)/2$ is the mean value of the conductivities at the two ends of the wire.]

Problem 170

A light bulb filament is constructed from 2 cm of tungsten wire of diameter 50 μm and is enclosed in an evacuated glass bulb. What temperature does the filament reach when it is operated at a power of 1 W? (Assume the emissivity of the tungsten surface to be 0.4.)

Solution

The power radiated by area A of a black body at temperature T is given by Stefan's law:

$$\frac{P}{A} = \sigma T^4,$$

where σ is the Stefan–Boltzmann constant. Thus the power per unit area radiated by a body of emissivity ε is $\varepsilon \sigma T^4$.

If the body's surroundings are at a temperature T_0, it will absorb a power per unit area of $\varepsilon \sigma T_0^4$ from them, so the net power emitted is

$$P = \varepsilon \sigma (T^4 - T_0^4) A.$$

Taking $\varepsilon = 0.4$, $\sigma = 5.67 \times 10^{-8}$ W m^{-2} K^{-4}, $P = 1$ W and $A = \pi \times 0.02 \times 50 \times 10^{-6}$ m$^2 = 3.14 \times 10^{-6}$ m^2 gives

$$T^4 - T_0^4 = 1.404 \times 10^{13} \text{ K}^4.$$

It is clear that the temperature T_0 of the surroundings can be ignored unless it is very high [it would have to be ≈ 900 K to make 1% difference to the answer], so we obtain $T = 1.94 \times 10^3$ K.

Problem 171

A small spherical satellite is in circular orbit round the Sun. The Sun subtends at the satellite a solid angle of 7×10^{-5} steradians, and the temperature of the satellite is uniform. Assuming that the emissivity of the satellite is independent of wavelength, calculate this temperature. (The effective black-body temperature of the Sun's surface is 5800 K.)

Solution

Let us write T for the satellite's temperature, r for its radius and ε for its emissivity. We will also write T_S for the Sun's effective temperature and Ω for the solid angle which it subtends at the satellite, and will temporarily introduce R for the Sun's radius and D for its distance from the satellite, as shown in figure 169. The relationship between Ω, R and D is

$$\Omega = \frac{\pi R^2}{D^2},$$

provided that $D \gg R$.

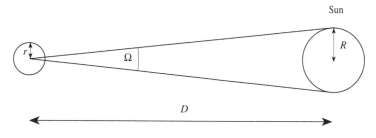

Figure 169

The total power emitted by the Sun in $4\pi R^2 \sigma T_S^4$, where σ is the Stefan–Boltzmann constant, so at a distance D the power per unit area is

$$\frac{4\pi R^2 \sigma T_S^4}{4\pi D^2} = \frac{\Omega \sigma T_S^4}{\pi}.$$

The satellite presents an area πr^2 to the incoming solar radiation, and since absorptivity and emissivity are equal, it must therefore absorb power

$$\pi r^2 \frac{\Omega \sigma T_S^4}{\pi} \varepsilon = r^2 \Omega \sigma T_S^4 \varepsilon.$$

Since the satellite has a surface area $4\pi r^2$ and a uniform temperature T it must radiate power

$$4\pi r^2 \sigma T^4 \varepsilon.$$

At equilibrium the absorbed and radiated powers must be equal, so

$$r^2 \Omega \sigma T_S^4 \varepsilon = 4\pi r^2 \sigma T^4 \varepsilon,$$

which can be rearranged to give

$$T = T_S \left(\frac{\Omega}{4\pi}\right)^{1/4}.$$

▶   Taking $\Omega = 7 \times 10^{-5}$ sr and $T_S = 5800$ K gives $T = 280$ K.
   [This problem gives a simplified understanding of the Earth's mean temperature, since the Sun subtends a mean solid angle of 6.80×10^{-5} sr at the Earth.]

Problem 172

A frictionless piston of mass m is a precise fit in the vertical cylindrical neck of a large container of volume V. The container is filled with a gas and there is a vacuum above the piston. The cross-sectional area of the neck is A.

   (a) Calculate the pressure of the gas in the container when the piston is in equilibrium.

   (b) Assuming that the pressure and volume of the gas are related by Boyle's law, calculate the restoring force on the piston when it is displaced by a small distance x.

   (c) Assuming that the motion of the piston is slow enough for Boyle's law to be valid, obtain the differential equation for small displacements of the piston about its equilibrium position.

   (d) Show that the angular frequency of oscillation ω is independent of m.

   (e) Calculate ω for $V = 2000$ litres and $A = 1.0 \times 10^{-4}$ m^2.

Solution

   (a) The weight of the piston is mg, and this must be balanced by the
▶   upward force pA exerted by the gas pressure, so $p = mg/A$.
   (b) Boyle's law states that pV is constant at constant temperature.

Differentiating this gives

$$p \, dV + V \, dp = 0.$$

If the piston is displaced upwards by a distance x, the increase dV in the volume of gas is Ax, so

$$dp = -\frac{p \, dV}{V} = -\frac{mg \, Ax}{A \ V} = -\frac{mg \, x}{V}.$$

▶   The restoring force on the piston is thus $-mgAx/V$, where the negative sign indicates that the force acts downwards.

(c) The acceleration of the piston is thus $-gAx/V$, which we can express as a differential equation:

▶
$$\frac{d^2x}{dt^2} = -\frac{gA}{V}x.$$

(d) We recognise this as the differential equation for simple harmonic motion, with an angular frequency of oscillation $\omega = (gA/V)^{1/2}$ which is independent of the mass m of the piston.

(e) If $V = 2000$ litres $= 2.0 \, \text{m}^3$, and $A = 1.0 \times 10^{-4} \, \text{m}^2$, we obtain

▶ $\omega = 2.2 \times 10^{-2} \, \text{s}^{-1}$. This corresponds to a period of oscillation of about 4.7 minutes.

Problem 173

A cylinder with adiabatic walls is closed at both ends and is divided into two volumes by a frictionless piston that is also thermally insulating. Initially, the volume, pressure and temperature of the ideal gas in each side of the cylinder are equal at V_0, p_0 and T_0 respectively. A heating coil in the right-hand volume is used to heat slowly the gas on that side until the pressure reaches $64p_0/27$. If the heat capacity C_v of the gas is independent of temperature, and $C_p/C_v = \gamma = 1.5$, find the following in terms of V_0, p_0 and T_0:

(a) the entropy change of the gas on the left;
(b) the final left-hand volume;
(c) the final left-hand temperature;
(d) the final right-hand temperature;
(e) the work done on the gas on the left.

Solution

Figure 170 shows the cylinder before and after the input of heat.

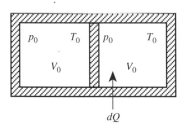
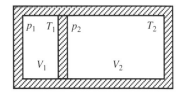

Figure 170

▶   (a) The piston is thermally insulating, so no heat can enter the left-hand side. Thus the entropy change of the gas on the left must be zero.

(b) Since the entropy of the gas on the left-hand side does not change, the compression must be adiabatic. Writing p_1 and V_1 for the pressure and volume of this gas after compression, we must have

$$p_0 V_0^\gamma = p_1 V_1^\gamma.$$

So

$$V_1 = V_0 \left(\frac{p_0}{p_1}\right)^{1/\gamma} = V_0 \left(\frac{p_0}{p_1}\right)^{2/3}.$$

Now at equilibrium, the pressures in the left- and right-hand sides must be equal, so that $p_1 = 64 p_0/27$. Thus the final volume of gas in the
▶ left-hand side must be $V_0(27/64)^{2/3} = 9V_0/16$.

(c) The number of molecules of gas in the left-hand side must also be constant during the compression. Since the ·umber of molecules is proportional to pV/T, we must have, writing T_1 for the final temperature,

$$\frac{p_0 V_0}{T_0} = \frac{p_1 V_1}{T_1}.$$

Thus

$$T_1 = \frac{p_1 V_1}{p_0 V_0} T_0,$$

so the final temperature of the gas on the left-hand side must be
▶ $(64/27)(9/16) T_0 = 4T_0/3$.

(d) The total volume of gas in both sides of the cylinder must be constant at $2V_0$, so the final volume in the right-hand side must be

$2V_0 - 9V_0/16 = 23V_0/16$. Let us call this V_2. If we put T_2 for the final temperature in the right-hand side, and recall that the final pressure there is $p_1 = 64p_0/27$, conservation of molecules on the right-hand side gives

$$\frac{p_0V_0}{T_0} = \frac{p_1V_2}{T_2}.$$

Thus

$$T_2 = \frac{p_1V_2}{p_0V_0}T_0,$$

▶ so $T_2 = (64/27)(23/16)T_0 = 92T_0/27$.

(e) No heat enters the gas on the left-hand side, so the work done on it must be equal to the increase in its internal energy ΔU, which equals $C_v\Delta T$. Since the specific heat C_v is independent of temperature, the total internal energy U must be C_vT, apart from an arbitrary constant which will not matter since we are only interested in energy differences.

Now we know that $C_p - C_v = NR$ (where N is the number of moles, and R is the gas constant) and $C_p/C_v = \gamma$, so $C_v = NR/(\gamma - 1)$. Thus $U = NRT/(\gamma - 1) = pV/(\gamma - 1)$, so we may finally write for the work done on the gas on the left-hand side

$$\frac{p_1V_1 - p_0V_0}{\gamma - 1} = \frac{\dfrac{64}{27}\dfrac{9}{16} - 1}{1/2}p_0V_0.$$

▶ This gives $2p_0V_0/3$ for the work done on the gas on the left-hand side.

Problem 174

The internal combustion petrol engine can be modelled on the Otto cycle. The four stages consist of (a) an *adiabatic* compression from V_1 to V_2, (b) an *isochoric* (constant volume) pressure increase, (c) an *adiabatic* expansion from V_2 to V_1 and (d) an *isochoric* pressure decrease. Sketch the $p - V$ diagram. Assuming that the gas behaves ideally, with a constant heat capacity, show that the efficiency is

$$\eta = 1 - \frac{1}{r^{\gamma - 1}},$$

where the compression ration $r = V_1/V_2$.

Solution

The p–V diagram is sketched in figure 171.

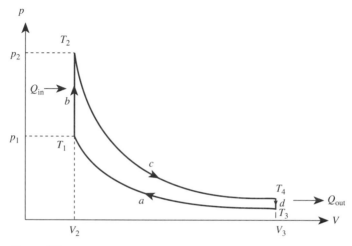

Figure 171

Let us assume there are N moles of gas present. We can use $pV = NRT$ to calculate the temperatures T_1 and T_2 at the ends of process (b). They are

$$T_1 = \frac{p_1 V_2}{NR} \text{ and } T_2 = \frac{p_2 V_2}{NR},$$

so the heat input during process (b) must be

$$Q_{\text{in}} = C_v N (T_2 - T_1) = \frac{C_v V_2}{R} (p_2 - p_1),$$

where C_v is the molar heat capacity at constant volume.

The processes (a) and (c) are adiabatic, so the relationship between V and T is

$$V^{\gamma - 1} T = \text{constant}.$$

The temperatures T_3 and T_4 at the ends of process (d) are thus

$$T_3 = \left(\frac{V_2}{V_1}\right)^{\gamma - 1} \frac{p_1 V_2}{NR} \text{ and } T_4 = \left(\frac{V_2}{V_1}\right)^{\gamma - 1} \frac{p_2 V_2}{NR}.$$

Thus the heat output during process (d) is

$$Q_{out} = C_v N(T_4 - T_3) = \left(\frac{V_2}{V_1}\right)^{\gamma-1} \frac{C_v V_2}{R}(p_2 - p_1) = \left(\frac{V_2}{V_1}\right)^{\gamma-1} Q_{in}.$$

I.e.

$$\frac{Q_{in}}{Q_{out}} = r^{\gamma-1}.$$

The work done, W, during the cycle, is equal to $Q_{in} - Q_{out}$, and the efficiency η is W/Q_{in}, so

$$\eta = \frac{Q_{in} - Q_{out}}{Q_{in}} = 1 - \frac{1}{r^{\gamma-1}}$$

as required.

Problem 175

A monatomic gas of molecular weight M is held at low pressure and temperature T. The velocity distribution function has the form

$$C \exp\left(-\beta[v_x^2 + v_y^2 + v_z^2]\right).$$

Write down the value of β, and derive the form of the energy distribution function. Sketch the form of this latter distribution.

Solution

The exponential term in the velocity distribution function is a Boltzmann factor, which has the general form $\exp(-E/kT)$, where E is the energy and k is Boltzmann's constant. Since the kinetic energy of a molecule of mass M and velocity (v_x, v_y, v_z) is $M(v_x^2 + v_y^2 + v_z^2)/2$, we can see that $\beta = M/2kT$.

To calculate the energy distribution $f(E)$, we consider a range of energies from E to $E + dE$ such that the fraction of molecules possessing kinetic energies within this range is defined to be $f(E)dE$. Now in velocity space, the set of points representing a constant energy E is a sphere of radius c centred on the origin, where c is the molecular speed corresponding to a kinetic energy E. Thus the region of velocity space corresponding to energies between E and $E + dE$ is a spherical shell of

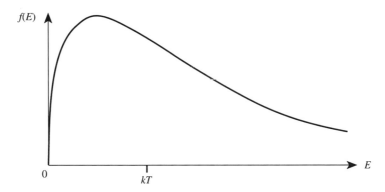

Figure 172

radius c and thickness dc, where

$$E = \frac{1}{2}Mc^2,$$

so

$$c = \left(\frac{2E}{M}\right)^{1/2} \text{ and } dc = \frac{1}{2}\left(\frac{2}{ME}\right)^{1/2}dE.$$

The value of the velocity distribution function is constant throughout this spherical shell (since its velocity dependence is in fact just a dependence on the speed c), so to calculate the total fraction of molecules $f(E)dE$ we merely need to multiply this constant value by the volume of the shell. Since the area of a sphere of radius c is $4\pi c^2$, the volume of the shell is $4\pi c^2 dc$, so we may write $f(E)dE$ as

$$4\pi c^2 dc \cdot C \exp(-\beta c^2).$$

Writing c and dc in terms of E gives

$$f(E)dE = 4\pi \frac{2E}{M}\frac{1}{2}\left(\frac{2}{ME}\right)^{1/2}dE \cdot C \exp\left(-\frac{2\beta E}{M}\right),$$

which can be written as

▶ $$f(E) = KE^{1/2}\exp\left(-\frac{E}{kT}\right),$$

where we have substituted our earlier expression for β and combined the factors which do not depend on E into the constant K. [This constant could be determined from the normalising condition that

$$\int_0^\infty f(E)dE = 1.$$

In fact, it is

$$K = \frac{1}{\sqrt{(2\pi)}} \left(\frac{2}{kT} \right)^{3/2} .]$$

The form of the energy distribution $f(E)$ is shown in figure 172. It has a maximum at $E = kT/2$.

Problem 176

Explain why the distribution of speeds of molecules emerging through a small hole in an effusive molecular beam source is not a Maxwellian distribution.

Solution

The short answer is that the faster molecules are more likely to emerge than the slower ones, so that the speed distribution will be weighted towards the higher speeds.

In more detail, following standard kinetic theory we can write

$$\frac{1}{2} n \sin \theta \, f(c) \, d\theta \, dc$$

for the number of molecules per unit volume in the bulk of the gas, having speeds between c and $c + dc$ and directions between θ and $\theta + d\theta$. n is the number of molecules per unit volume, and $f(c)$ is the Maxwell speed distribution.

If we now consider the number of molecules passing through an area A of the surface in unit time, it is clear that molecules of speed c must travel a distance of at most c if they are to escape in unit time. This is shown in figure 173.

The volume of space occupied by these molecules, travelling at an angle θ to the surface normal, is

$$cA \cos \theta,$$

so the number of molecules emerging per unit time with speeds in the range c to $c + dc$ and directions in the range θ to $\theta + d\theta$ is

$$\frac{1}{2} n \cos \theta \sin \theta \, c f(c) \, d\theta \, dc,$$

▶   i.e. the speed distribution is $c f(c)$ which is not Maxwellian.

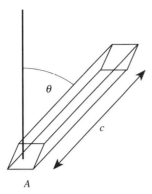

Figure 173

Problem 177

The distribution of molecular velocities may be written

$$P(v_x)dv_x = \left(\frac{m}{2\pi kT}\right)^{1/2} \exp\left(-\frac{mv_x^2}{2kT}\right)dv_x,$$

where $P(v_x)dv_x$ is the probability that a molecule will have a velocity component (in the x-direction) in the range v_x to $v_x + dv_x$, m is the molecular mass, T the temperature and k Boltzmann's constant.

From this derive the Maxwell distribution in molecular speeds (in all directions). Sketch the form of both distributions.

A vessel contains a monatomic gas at temperature T. Use Maxwell's distribution to calculate the mean kinetic energy of the molecules.

Molecules of the gas stream through a small hole into a vacuum. A box is opened for a short time and catches some of the molecules. What will be the final temperature of the gas trapped in the box? (The thermal capacity of the box is to be ignored.)

If

$$I_n = \int_0^\infty x^n \exp(-ax^2)dx$$

then

$$I_n = \frac{n-1}{2a}I_{n-2}$$

and

$$I_0 = \frac{1}{2}\sqrt{\frac{\pi}{a}}, \; I_1 = \frac{1}{2a}.)$$

Solution

The first part of this problem is very similar to problem 175, in which we calculated the energy distribution function. As before, we note that the region of velocity space corresponding to speeds between c and $c + dc$ is a spherical shell of radius c and thickness dc. This shell has a volume of $4\pi c^2 dc$. The three-dimensional velocity distribution function is given by

$$p(v_x, v_y, v_z)\,dv_x dv_y dv_z = P(v_x)P(v_y)P(v_z)\,dv_x dv_y dv_z,$$

where $p(v_x, v_y, v_z)\,dv_x dv_y dv_z$ is the probability that the x-component of the velocity is between v_x and $v_x + dv_x$, the y-component is between v_y and $v_y + dv_y$, and the z-component is between v_z and $v_z + dv_z$. The value of $p(v_x, v_y, v_z)$ is constant throughout the shell, so the probability that the speed is between c and $c + dc$ is given by the product of this value and the volume of the shell. Thus

▶
$$f(c)\,dc = \left(\frac{m}{2\pi kT}\right)^{3/2} \exp\left(-\frac{mc^2}{2kT}\right) 4\pi c^2\,dc.$$

The shapes of these distributions are shown in figure 174.

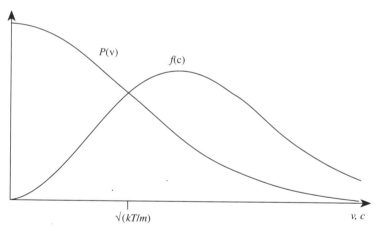

Figure 174

Since the kinetic energy E of a molecule is given by

$$E = \frac{1}{2}mc^2,$$

the mean kinetic energy $\langle E \rangle$ must be given by

$$\langle E \rangle = \frac{1}{2}m\int_0^\infty c^2 f(c)\,dc.$$

We have just shown that $f(c)$ is given by $Ac^2 \exp(-mc^2/2kT)$ where A is a constant for molecules of a particular mass at a particular temperature. Although we know the value of A, it is perhaps easier to note that $\int f(c)dc = 1$, so that

$$A = \frac{1}{\displaystyle\int_0^\infty c^2 \exp(-mc^2/2kT)dc}.$$

Thus

$$\langle E \rangle = \frac{1}{2}m\, \frac{\displaystyle\int_0^\infty c^4 \exp(-mc^2/2kT)\, dc}{\displaystyle\int_0^\infty c^2 \exp(-mc^2/2kT)\, dc}.$$

We recognise this expression as having the form

$$\frac{1}{2}m\frac{I_4}{I_2},$$

▶ where $a = m/2kT$. Now $I_4/I_2 = 3/2a = 3kT/m$, so $\langle E \rangle = 3kT/2$.

In problem 176 we showed that the molecules which emerge from a small hole have a speed distribution proportional to $c f(c)$. The mean kinetic energy of the emerging molecules is thus given by

$$\langle E \rangle = \frac{1}{2}m\, \frac{\displaystyle\int_0^\infty c^2 c\, f(c)\, dc}{\displaystyle\int_0^\infty c\, f(c)\, dc}.$$

In a similar manner to our calculation of the mean kinetic energy of the molecules in the bulk of the gas, we can write this as

$$\langle E \rangle = \frac{1}{2}m\, \frac{\displaystyle\int_0^\infty c^5 \exp(-mc^2/2kT)\, dc}{\displaystyle\int_0^\infty c^3 \exp(-mc^2/2kT)\, dc},$$

i.e. $mI_5/2I_3$. Since $I_5/I_3 = 2/a = 4kT/m$, this gives $\langle E \rangle = 2kT$. Thus the molecules which emerge from the gas have kinetic energy greater by $kT/2$, on average, than the molecules in the bulk of the gas. When these molecules are trapped and allowed to come into equilibrium, they will reach a temperature T' such that $3kT'/2 = 2kT$, and hence the final
▶ temperature will be $4T/3$.

Problem 178

A thin disc of radius R is suspended in a gas whose molecular mean free path greatly exceeds the diameter of the disc. The disc moves very slowly through the gas with velocity W in a direction normal to its surface. Derive an expression for the force resisting the motion in terms of R, W, n, m and $\langle c \rangle$, where n is the number of molecules per unit volume, $\langle c \rangle$ is their mean speed and m is the mass of each.

Solution

Let us start by considering a single molecule approaching the disc at speed c and in a direction which makes an angle θ with the normal to the disc, as shown in figure 175.

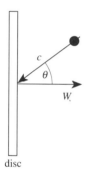

disc

Figure 175

The component of the molecule's velocity normal to the disc is $c \cos \theta$, so the normal component of its relative velocity is $c \cos \theta + W$. If the mass of the molecule is very much less than that of the disc (a reasonable assumption!), the disc's motion can be assumed to be unchanged as a result of the collision, and the molecule will rebound such that the normal component of its relative velocity is reversed. (The parallel component will be unchanged since the disc exerts no force parallel to its surface.) The change in the molecule's velocity is thus $2(c \cos \theta + W)$ to the right, so the impulse (change in momentum) exerted on the disc as a result of this single collision is

$$2m(c \cos \theta + W).$$

In order to calculate a force from this, we need to know the rate at which such impacts occur. This will depend on the speed c and direction θ of the molecules, so we will need to integrate over the velocity distribution.

In unit time, the disc moves a distance W to the right, and molecules with speed c move a distance c in the direction θ. The molecules which can collide with the disc are thus contained in the shaded region of figure 176.

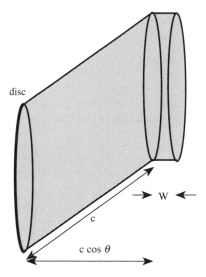

Figure 176

This region has a volume $(c \cos \theta + W)\pi R^2$, so it contains $n\pi R^2(c \cos \theta + W)$ molecules. We know that the fraction of molecules having speeds between c and $c + dc$ and directions between θ and $\theta + d\theta$ is

$$\frac{1}{2}f(c) \sin \theta \, dc \, d\theta,$$

so the number of collisions per unit time arising from molecules having speeds between c and $c + dc$ and directions between θ and $\theta + d\theta$ is

$$\frac{1}{2}n\pi R^2(c \cos \theta + W)f(c) \sin \theta \, dc \, d\theta.$$

Each of these molecules transfers momentum $2m(c \cos \theta + W)$ to the disc, so the force exerted on this side of the disc is

$$mn\pi R^2(c \cos \theta + W)^2 f(c) \sin \theta \, dc \, d\theta.$$

We can now integrate this from $c = 0$ to ∞ and from $\theta = 0$ to $\pi/2$ (we are still dealing with only one side of the disc) to find the total force acting on this side:

$$F = mn\pi R^2 \int_{c=0}^{\infty} \int_{\theta=0}^{\pi/2} (c \cos \theta + W)^2 f(c) \sin \theta \, dc \, d\theta$$

$$= mn\pi R^2 \left(\int_0^{\infty} c^2 f(c) \, dc \int_0^{\pi/2} \cos^2 \theta \sin \theta \, d\theta \right.$$

$$\left. + 2W \int_0^{\infty} c f(c) \, dc \int_0^{\pi/2} \cos \theta \sin \theta \, d\theta + W^2 \int_0^{\pi/2} \sin \theta \, d\theta \right).$$

Now $\int \cos^2 \theta \sin \theta \, d\theta$ evaluated from $\theta = 0$ to $\theta = \pi/2$ is $1/3$, and $\int c^2 f(c) \, dc$ evaluated from $c = 0$ to $c = \infty$ is, by definition $\langle c^2 \rangle$, so the first term in the bracket has a value of $\langle c^2 \rangle /3$. Similarly, $\int \cos \theta \sin \theta \, d\theta = 1/2$ and $\int c f(c) \, dc = \langle c \rangle$, so the second term in the bracket has a value of $W \langle c \rangle$. Finally, $\int \sin \theta \, d\theta = 1$ so the third term in the bracket has a value of W^2. Thus the force acting on this side of the disc is

$$mn\pi R^2 \left(\frac{\langle c^2 \rangle}{3} + W \langle c \rangle + W^2 \right).$$

[We note in passing that this gives the correct result $mn \langle c^2 \rangle /3$ for the pressure on a stationary object.] The total force acting on the other side of the disc can be found by replacing W with $-W$ (since the other side of the disc is retreating from the molecules rather than advancing towards them), i.e.

$$mn\pi R^2 \left(\frac{\langle c^2 \rangle}{3} - W \langle c \rangle + W^2 \right),$$

and since these forces act in opposite directions the net force on the disc is the difference between them, giving our final answer as $2mn\pi R^2 W \langle c \rangle$.

Problem 179

A chamber contains oxygen at a pressure of $120 \, \mathrm{N \, m^{-2}}$ and at room temperature. Estimate the time taken for 1% of a clean surface inside the chamber to become coated with a single layer of oxygen molecules, assuming that all the molecules which strike the surface stick to it. (The effective diameter of the oxygen molecule is 0.3 nm.)

Solution

If the number of molecules per unit volume in the chamber is n, and their mean speed is $\langle c \rangle$, the number of molecules arriving per unit area of surface per unit time is

$$n \langle c \rangle / 4.$$

Since these molecules stick to the surface, the rate at which the fraction f of the surface covered is increasing is

$$\frac{df}{dt} = \frac{n \langle c \rangle}{4} \frac{\pi d^2}{4} = \frac{n \pi d^2 \langle c \rangle}{16},$$

where d is the molecular diameter.

Now

$$n = \frac{p}{kT}$$

and

$$\langle c \rangle = \sqrt{\frac{8kT}{\pi m}},$$

where p is the pressure, T is the temperature, k is Boltzmann's constant and m is the molecular mass, so

$$\frac{df}{dt} = \frac{\sqrt{(2\pi)}}{8} \frac{pd^2}{\sqrt{(mkT)}}.$$

Taking $m = 32 \times 1.66 \times 10^{-27}$ kg $= 5.31 \times 10^{-26}$ kg and $T = 293$ K, and substituting the values given for p and T, gives $df/dt = 2.31 \times 10^5$ s^{-1}. Thus the time required for f to increase from zero to 0.01 is

▶ $0.01/(2.31 \times 10^5)$ s $= 4 \times 10^{-8}$ s.

[In fact, the value of df/dt that we have calculated is only valid as long as f remains small. When an appreciable fraction of the surface has been covered, df/dt will be smaller because some of the molecules that arrive will hit parts of the surface that are already covered. However, since we are considering $f < 0.01$, the correction needed to take account of this effect will be negligible.]

Problem 180

A vacuum pump, which maintains a pressure of 10^{-7} Pa, is connected to a large vacuum chamber, of volume 1 m^3, by a flange with a circular aperture of radius 10^{-2} m. The chamber contains air at an initial pressure of 10^{-5} Pa. How long does it take for the pressure to fall to 2×10^{-7} Pa?

(You may assume a value of $450 \, \text{m s}^{-1}$ for $\langle c \rangle$ and that the mean free path of air molecules at atmospheric pressure is 100 nm.)

Solution

The mean free path at constant temperature is inversely proportional to the pressure, so at 10^{-5} Pa it must be $100 \, \text{nm} \times (10^5/10^{-5}) = 1000$ m. It is thus much larger than the aperture, so we are justified in using the formula

$$\frac{1}{4} n \langle c \rangle A$$

for the number of molecules emerging in unit time from an aperture of area A. n is the number of molecules per unit volume in the vessel from which the molecules are escaping, and $\langle c \rangle$ is their mean speed.

Put n_c for the number of molecules per unit volume in the vacuum chamber, and n_p for the number of molecules per unit volume in the pump. The rate at which molecules leave the chamber is thus

$$\frac{1}{4} n_c \langle c \rangle A$$

and the rate at which they enter it from the pump is

$$\frac{1}{4} n_p \langle c \rangle A.$$

If the volume of the vacuum chamber is V, we can therefore write

$$\frac{dn_c}{dt} = \frac{1}{V} \left(\frac{1}{4} n_p \langle c \rangle A - \frac{1}{4} n_c \langle c \rangle A \right).$$

Now for a particular gas at a constant temperature, the number of molecules per unit volume is proportional to the pressure, so we can rewrite this as

$$\frac{dp_c}{dt} = \frac{\langle c \rangle A}{4V} (p_p - p_c).$$

This differential equation can most easily be solved by noting that p_p is constant and putting

$$p' = p_c - p_p,$$

whereupon the equation becomes

$$\frac{dp'}{dt} = -\frac{\langle c \rangle A}{4V} p'.$$

The solution is thus

$$p' = B \exp(-\langle c \rangle At/4V),$$

where B is a constant. If we now put p_i for the initial value of p_c, the constant B must be $p_i - p_p$, and our expression for the variation of p_c with time becomes

$$p_c = p_p + (p_i - p_p) \exp(-\langle c \rangle At/4V).$$

Rearranging this to make time the subject gives

$$t = \frac{4V}{\langle c \rangle A} \ln \frac{p_i - p_p}{p_c - p_p}.$$

▶   Substituting the values given in the problem gives $t = 130$ s.

Problem 181

A gas containing equal number densities of two isotopes having masses of $349 \, m_u$ and $352 \, m_u$ is passed through a porous membrane in which the pores are much smaller than the mean free path. The gas emerges at much lower pressure and is pumped away. Show that the emergent gas is richer in one isotope. How many successive passages through similar membranes would be needed to produce a pure isotope contaminated by less than 1%?

Solution

Since the pores are much smaller than the mean free path, the formula

$$\frac{1}{4} n \langle c \rangle$$

can be used for the rate of molecular efflux per unit area. Thus the rate at which gas molecules pass through the membrane will be proportional to $n \langle c \rangle$.

At a constant temperature, the mean speed $\langle c \rangle$ is proportional to $m^{-1/2}$ where m is the molecular mass. Since the two isotopes initially have equal values of n, the lighter molecular species will emerge from the membrane more rapidly than the heavier one, and the emergent gas will be richer in the lighter isotope.

After the first pass through the membrane, the collected gas will have

$$\frac{n_{349}}{n_{352}} = \sqrt{\frac{352}{349}}$$

where n_{349} is the number density of the lighter isotope. When this mixture is passed through the membrane, the ratio will become $352/349$ (a factor of $\sqrt{(352/349)}$ comes from the ratio of the values of n, and an identical factor from the ratio of the values of $\langle c \rangle$). Thus after N passes through the membrane, the ratio becomes

$$\frac{n_{349}}{n_{352}} = \left(\frac{352}{349}\right)^{N/2}.$$

If we require the resultant mixture to consist of the lighter isotope contaminated with at most 1% of the heavier isotope, we must have

$$\frac{n_{349}}{n_{352}} > 99.$$

Thus

$$\frac{N}{2} \log \frac{352}{349} > \log 99,$$

▶   from which the required number of passes is $N \geqslant 1074$.

Problem 182

A vacuum flask is constructed with two concentric glass cylinders with a gap of 5 mm between the two which is evacuated to a pressure p. How low must p be before the thermal conductance between the inner and outer walls of the flask is reduced below its value when p is equal to 1 atmosphere? What value of p is required to reduce the thermal conductance to 10^{-3} of this value?

(You may assume the mean free path of air molecules to be 100 nm at atmospheric pressure, which can be taken to be $10^5 \, \text{N m}^{-2}$.)

Solution

The thermal conductivity of an ideal gas can be shown by transport arguments to be given by

$$K = \frac{1}{3}nl\langle c \rangle C_v',$$

where

> n is the number density of molecules,
> l is the mean free path,
> $\langle c \rangle$ is the mean molecular speed, and
> C_v' is the molecular heat capacity.

Now the pressure p of the gas is given by

$$p = nkT,$$

where k is Boltzmann's constant and T is the temperature, and the mean free path is given approximately by

$$l = \frac{1}{n\sigma},$$

where σ is the molecular cross-section, so we expect the thermal conductivity to be independent of pressure at constant temperature. However, when the pressure is reduced below the value at which the mean free path becomes 5 mm, the effective mean free path will remain at 5 mm since the predominant mechanism for the transport of heat between the walls of the vacuum flask will be collisions of molecules with the walls. Since the mean free path is inversely proportional to the pressure at constant temperature, this will occur when the pressure is reduced below

$$\frac{100 \text{ nm}}{5 \text{ mm}} \times 10^5 \text{ Pa},$$

▶ i.e. 2 Pa.

Below this pressure, the thermal conductivity will be inversely proportional to pressure, so it will be reduced to 10^{-3} of its value at
▶ 1 atmosphere when the pressure is reduced to 2×10^{-3} Pa.

Problem 183

A column of water contains fine metal particles of radius 2×10^{-8} m, which are in thermal equilibrium at 25 °C. If there are 1000 particles per unit volume at a given height, how many particles would be found in the same volume 1 mm higher? The density of the metal is $2 \times 10^4 \text{ kg m}^{-3}$.

Solution

We know that the number density $n(x)$ will be proportional to

$$\exp(-E/kT),$$

where E is the potential energy of a particle at height x. If a particle has volume V, its weight is $\rho g V$ where ρ is the density of the metal and g is the gravitational field strength. However, the particle will also experience an upthrust of $\rho_0 g V$, where ρ_0 is the density of water, so that the net downward force is $(\rho - \rho_0)gV$. The increase in potential energy E for a rise in height x is thus $(\rho - \rho_0)gVx$. Taking

$$V = \frac{4\pi r^3}{3},$$

we can thus calculate the increase in potential energy for a rise of 1 mm:

$$\Delta E = 10^{-3} \times (2 \times 10^4 - 10^3) \times 9.81 \times \frac{4\pi}{3} \times (2 \times 10^{-8})^3 \text{ J}$$

$$= 6.25 \times 10^{-21} \text{ J}.$$

The concentration of particles therefore decreases by a factor of $\exp(\Delta E/kT) = 4.57$, so the concentration will be $1000/4.57 =$ 220 particles per unit volume. Note that we would have made an error of about 10% if we had neglected the upthrust.

Problem 184

Estimate the height at which atmospheric pressure is half the value at sea level, making clear your assumptions.

Solution

If air behaves like an ideal gas, and the temperature T of the atmosphere is constant, the pressure will be proportional to the density, and the density will be proportional to

$$\exp(-mgz/kT),$$

where m is the mass of an 'air molecule', g is the gravitational field strength (assumed constant), z is the height, and k is Boltzmann's constant. The height at which the pressure is halved is thus given by

$$\frac{mgz}{kT} = \ln 2,$$

so

$$z = \frac{kT}{mg} \ln 2.$$

Since air consists of about 80% N_2 and 20% O_2 we may take m to be $(0.8 \times 28 + 0.2 \times 32) \times 1.66 \times 10^{-27}$ kg $= 4.78 \times 10^{-26}$ kg. A suitable value for T might be 283 K (10 °C), $k = 1.38 \times 10^{-23}$ J K^{-1} and
▶ $g = 9.81$ m s^{-2}, so $z = 5.8$ km.

[The main error in this calculation is the assumption that the atmospheric temperature T is constant. In fact, it decreases more or less linearly with height, at about 0.0065 K m^{-1}, for the first 11 km. We can allow for this as follows.

For a stationary fluid in a gravitational field g, the variation of pressure p with height z is given by the *hydrostatic equation*

$$\frac{dp}{dz} = -\rho g,$$

where ρ is the density. For an ideal gas consisting of molecules of mass m, the relationship between pressure and density can be found from

$$pV = NkT,$$

where V is the volume occupied by N molecules of the gas. The mass of these N molecules is clearly $Nm = pVm/kT$, so the density ρ is pm/kT. Substituting this expression for ρ into the differential equation for p gives

▶ $$\frac{dp}{dz} = -\frac{mgp}{kT},$$

which can be rearranged to give

$$\frac{dp}{p} = -\frac{mg \, dz}{kT}.$$

Integrating this differential equation with T constant would give us back our exponential expression for the variation of pressure with height. However, let us instead put

$$T = T_0 - \alpha z$$

to obtain

$$\int_{p_0}^{p} \frac{dp'}{p'} = -\frac{mg}{k} \int_0^z \frac{dz'}{T_0 - \alpha z'}.$$

This expression can be integrated to give

$$\ln\left(\frac{p}{p_0}\right) = -\frac{mg}{\alpha k}\ln\left(\frac{T_0}{T_0 - \alpha z}\right),$$

which, on exponentiating, becomes

$$\frac{p}{p_0} = (1 - \alpha z/T_0)^{mg/\alpha k}.$$

If we take $m = 4.78 \times 10^{-26}$ kg as before, and $\alpha = 0.0065$ K m^{-1}, we obtain $mg/\alpha k = 5.23$. Thus

$$(1 - \alpha z/T_0) = 0.5^{1/5.23} = 0.876,$$

so

$$\alpha z/T_0 = 0.124.$$

▶ Taking $T_0 = 288$ K (the standard sea-level value) gives $z = 5.5$ km.]

Problem 185

A closed cylinder of radius a contains gas at constant temperature T with molecules each of mass m. The cylinder is made to rotate about its axis with angular velocity ω. Use the Boltzmann distribution law to derive the variation of density for the gas with distance from the axis. At what distance from the axis is the density unchanged by the rotation, if ω is small?

Solution

For definiteness, let us call the length of the cylinder L and the initial density of the gas (when the cylinder is not rotating) ρ_0. The total mass of gas within the cylinder, which must be constant since the cylinder is closed, is therefore

$$M = \pi a^2 L \rho_0.$$

When the cylinder is rotated, it becomes energetically favourable for the molecules to move towards the edge of the cylinder, so we expect the density to increase with radius r. Thermal excitation will prevent all of the molecules from being pushed to the edge of the cylinder.

In a frame of reference rotating with the gas, a molecule of mass m at a distance r from the axis experiences a centrifugal force

$$m\omega^2 r$$

away from the axis. By integrating this force with respect to r, we can derive the *centrifugal potential energy*

$$V = -\frac{1}{2}m\omega^2 r^2.$$

From the Boltzmann distribution law we expect the probability of a molecule being located at distance r, and hence the density at distance r, to be proportional to $\exp(-V/kT)$ where k is Boltzmann's constant and T is the temperature. Thus we can write the density as

$$\rho = A\exp(m\omega^2 r^2/2kT).$$

The constant A must be chosen to keep the mass of gas within the cylinder constant.

Clearly the total mass of gas within the cylinder is given by

$$M = 2\pi L \int_0^a \rho\, r\, dr = 2\pi L A \int_0^a r \exp(m\omega^2 r^2/2kT)\, dr.$$

This integral can be evaluated by noting that

$$\frac{d}{dr}\exp(m\omega^2 r^2/2kT) = \frac{m\omega^2 r}{kT}\exp(m\omega^2 r^2/2kT),$$

so that

$$M = \frac{2\pi L A kT}{m\omega^2}(\exp[m\omega^2 a^2/2kT] - 1).$$

Combining this with our expression for ρ, and our value for the total mass M, gives

▶
$$\rho = \frac{\rho_0 m\omega^2 a^2 \exp(m\omega^2 r^2/2kT)}{2kT\,(\exp[m\omega^2 a^2/2kT] - 1)}.$$

To simplify this, let us put

$$R = \frac{r}{a},\ \Omega = \frac{m\omega^2 a^2}{2kT}.$$

We may then write

$$\frac{\rho}{\rho_0} = \frac{\Omega\exp(\Omega R^2)}{\exp(\Omega) - 1}.$$

This function is sketched in figure 177, for different values of Ω.

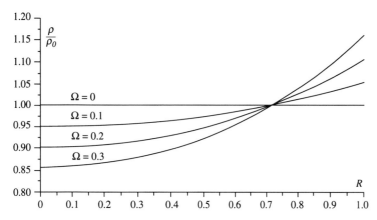

Figure 177

Writing the function as a series expansion for small Ω, we have

$$\frac{\rho}{\rho_0} \approx \frac{\Omega(1 + \Omega R^2 + \ldots)}{\Omega + \Omega^2/2 + \ldots} = \frac{\Omega(1 + \Omega R^2 + \ldots)}{\Omega(1 + \Omega/2 + \ldots)}.$$

▶ For small Ω this is clearly equal to 1 when $R^2 = 1/2$. Thus the density is unchanged for small angular velocity at a distance $r = a/\sqrt{2}$ from the axis.

Problem 186

A problem with high temperature metal oxide superconductors, which are prepared at 900 °C, is that the oxygen content is too low. This problem can be overcome by diffusing oxygen in from the surface at 400 °C.

How long will this process take in a crystal of smallest dimension 10^{-3} m, and how long does it take to lower the oxygen content again by heating at 900 °C? Assume that the energy barrier that must be surmounted to move an oxygen atom from one site to another is 1 eV, that the intersite distance is 2×10^{-10} m, and that the characteristic frequency of vibration of the lattice is 10^{14} Hz. (You may choose to consider this problem as an example of a one-dimensional random walk.)

Solution

Write W for the height of the energy barrier and T for the (thermodynamic) temperature. The probability p of an oxygen atom surmounting this barrier and thus moving to an adjacent site, at a single attempt, is

$$p = \exp(-W/kT),$$

where k is Boltzmann's constant. The expectation number of attempts required before a jump occurs is* $1/p$, so if we write v for the characteristic lattice vibration frequency, the mean time between jumps from one site to a neighbouring site will be given by

$$\langle t \rangle = \frac{1}{vp} = \frac{\exp(W/kT)}{v}.$$

We can view the diffusion process as a one-dimensional random walk in which steps of length a (the intersite distance) are taken at time intervals $\langle t \rangle$. It is a well-known result that for a one-dimensional random walk, the root mean square distance from the starting point is $\sqrt{N}$ step lengths after N steps. Thus after time t, an oxygen atom will have taken $t/\langle t \rangle$ steps and its root mean square distance x from its starting point will be given by

$$x = a\sqrt{\frac{t}{\langle t \rangle}}.$$

Thus the time required for a significant number of oxygen atoms to diffuse a distance x through the crystal is

$$t = \frac{x^2}{a^2}\langle t \rangle = \frac{x^2}{a^2 v}\exp(W/kT).$$

From the data given in the problem, we have

$W = 1\,\text{eV} = 1.6 \times 10^{-19}\,\text{J},$
$a = 2 \times 10^{-10}\,\text{m},$
$T = 400°\text{C} = 673\,\text{K},$
$v = 10^{14}\,\text{Hz},$
$x = 5 \times 10^{-4}\,\text{m}$ (since oxygen atoms can diffuse in from either side of the crystal).

Substituting these values into our expression for t, we find

▶ $t = 1.9 \times 10^6\,\text{s} \approx 22$ days.

The time taken for the atoms to diffuse out again at 900 °C can be

▶ found by recalculating with all values kept the same except for T, which is changed to 1173 K. This gives $t = 1.2 \times 10^3$ s ≈ 20 minutes.

[* I am not sure if this is obvious, though it is clearly reasonable. We can prove it as follows. If the probability of a successful jump is p, the probability that the first successful jump occurs at the nth attempt must be $(1 - p)^{n-1}p$ (representing $n - 1$ failures followed by one success). The expectation value of the number of attempts is therefore

$$\langle n \rangle = p \sum_{j=1}^{\infty} j(1 - p)^{j-1}.$$

If we put $x = 1 - p$, the sum becomes $\langle n \rangle = p(1 + 2x + 3x^2 + 4x^3 + \ldots)$, which looks familiar. If we define a new sum $S(x) = 1 + x + x^2 + x^3 + \ldots$ and differentiate it with respect to x, it is clear that

$$\langle n \rangle = p\frac{dS}{dx}.$$

Now S is just a geometric progression, with a sum

$$S(x) = \frac{1}{1 - x},$$

so

$$\frac{dS}{dx} = \frac{1}{(1 - x)^2}.$$

Thus $\langle n \rangle = p/(1 - x)^2 = p/p^2 = 1/p$.]

Problem 187

A paramagnetic solid owes its magnetic properties to the fact that some (N per unit volume) of the atoms have permanent magnetic dipole moments (μ per atom). In the presence of a magnetic flux density B each moment can be regarded as being allowed to exist in only two states, parallel or antiparallel to the field, with energies $-\mu B$ and $+\mu B$ respectively. Obtain an expression for the fraction of moments that point parallel to the field at temperature T, and hence calculate the magnetic susceptibility of the solid.

The solid is held at a temperature of 1 K in a field of 1 T. The solid is then thermally isolated and the field is reduced to 0.3 T. If no dipoles change their orientation during the reduction of the field, to what temperature does the final distribution of the dipoles correspond?

Solution

The Boltzmann distribution tells us that the probabilities of a dipole being oriented parallel and antiparallel to the field are

$$p_{\text{parallel}} = A \exp\left(\mu B / kT\right),$$
$$p_{\text{antiparallel}} = A \exp\left(-\mu B / kT\right).$$

The normalising constant A can be found by realising that no other possibility exists, so that $p_{\text{parallel}} + p_{\text{antiparallel}} = 1$. Thus

$$A = \frac{1}{\exp\left(\mu B / kT\right) + \exp\left(-\mu B / kT\right)}.$$

The fraction f of moments that are parallel to the field is equal to p_{parallel}, so

▶
$$f = \frac{\exp\left(\mu B / kT\right)}{\exp\left(\mu B / kT\right) + \exp\left(-\mu B / kT\right)}.$$

A unit volume of the solid contains N dipoles, of which fN are oriented parallel to the field and $(1 - f)N$ antiparallel. The net dipole moment per unit volume, M, is thus equal to

$$M = \mu(fN - [1 - f]N) = \mu N(2f - 1).$$

Inserting our expressions for f, this gives

$$M = \mu N \frac{\exp\left(\mu B / kT\right) - \exp\left(-\mu B / kT\right)}{\exp\left(\mu B / kT\right) + \exp\left(-\mu B / kT\right)} = \mu N \tanh\left(\frac{\mu B}{kT}\right).$$

The susceptibility χ is given by

$$1 + \chi = \frac{B}{\mu_0 H} = \frac{B}{B - \mu_0 M},$$

so

▶
$$\chi = \frac{\mu_0 M}{B - \mu_0 M} = \frac{\mu_0 \mu N \tanh\left(\mu B / kT\right)}{B - \mu_0 \mu N \tanh\left(\mu B / kT\right)}.$$

[At sufficiently high temperatures and low magnetic fields, this tends to

$$\frac{\mu_0 \mu^2 N}{kT}.$$

This proportionality of χ to $1/T$ is called the *Curie law*. The condition that the temperature is high enough is that $\mu B / kT \ll 1$. Since μ is usually of the order of a *Bohr magneton* ($9.3 \times 10^{-24}\ \text{J T}^{-1}$), the temperature need only be greater than about 1 K for fields as large as 1 T.]

When the solid is in thermal equilibrium at temperature T and in a magnetic field B, the ratio of dipoles oriented parallel to the field to those oriented antiparallel is $\exp(2\mu B/kT)$. As the magnetic field is reduced, this ratio is held constant which implies that the corresponding temperature is proportional to the magnetic field. Thus if the field is reduced from 1 T to 0.3 T, the corresponding temperature must fall from ▶ 1 K to 0.3 K. [This demonstrates the principle of cooling by adiabatic demagnetisation.]

Problem 188

The equilibrium separation between hydrogen atoms in the hydrogen molecule is 0.08 nm and the force constant of the bond is $580\,\mathrm{N\,m^{-1}}$. Estimate the energy needed to excite (a) the lowest rotational mode, and (b) the lowest vibrational mode.

Sketch and comment on the variation of specific heat capacity of hydrogen gas between 30 K and 1000 K.

Solution

(a) Rotational mode.
We can model the molecule as a dumb-bell consisting of two masses m (the mass of a hydrogen atom) separated by a distance d, as shown in figure 178.

Figure 178

The moment of inertia I of the molecule is

$$I = 2m\left(\frac{d}{2}\right)^2$$
$$= \frac{md^2}{2}.$$

The rotational energy, according to quantum mechanics, is

$$E_{rot} = \frac{J(J+1)h^2}{8\pi^2 I},$$

where h is Planck's constant and J is the rotational quantum number, which must be an integer, so the lowest non-zero rotational energy is

$$E_{min} = \frac{h^2}{4\pi^2 I} = \frac{h^2}{2\pi^2 md^2}.$$

Substituting $h = 6.626 \times 10^{-34}$ J s, $m = 1.67 \times 10^{-27}$ kg and ▶ $d = 8 \times 10^{-11}$ m gives $E_{min} = 2.1 \times 10^{-21}$ J.
(b) Vibrational mode.
We now model the molecule as two masses m connected by a spring of spring constant k. The masses vibrate along the molecular axis, as shown in figure 179.

Figure 179

The vibrational frequency v of this arrangement is

$$v = \frac{1}{2\pi}\sqrt{\frac{2k}{m}},$$

so the energy needed to excite the lowest vibrational mode is

$$E_{min} = hv = \frac{h}{2\pi}\sqrt{\frac{2k}{m}}.$$

[Actually, the quantum-mechanical analysis shows that the vibrational energy is given by $E = (n + 1/2)hv$ where n is an integer. The lowest mode has $n = 0$ and an energy $hv/2$ (the zero-point energy), but the energy required to excite the mode $n = 1$ is greater than this by hv so our analysis is still valid.]
▶   Substituting the values for k and m gives $E_{min} = 8.8 \times 10^{-20}$ J.
   The temperatures at which the lowest modes will be excited are given approximately by

$$\frac{E_{min}}{k},$$

where k is Boltzmann's constant, so the lowest rotational mode will be excited at about 150 K and the lowest vibrational mode at about 6000 K. Thus at temperatures below about 150 K, neither vibrational modes nor rotational modes will be excited and the only way in which the molecule can store thermal energy is as kinetic energy of translation. It will therefore have three degrees of freedom, so the ratio of specific heats, γ, will be $1 + 2/3 = 5/3$. The molar heat capacity at constant volume, C_v, is given by

$$C_v = \frac{R}{\gamma - 1},$$

where R is the molar gas constant, so at temperatures below about 150 K we expect hydrogen to have a molar heat capacity at contant volume of about $3R/2 = 12.5 \, \mathrm{J \, mol^{-1} \, K^{-1}}$. We can convert this to the specific heat capacity by dividing by the molar mass of hydrogen, which is $0.002 \, \mathrm{kg \, mol^{-1}}$, to give $6.25 \, \mathrm{kJ \, kg^{-1} \, K^{-1}}$.

Between about 150 K and 6000 K, the rotational mode will also be excited, contributing a further two degrees of freedom. The ratio of specific heats will become $\gamma = 1 + 2/5 = 7/5$, and C_v will become $5R/2 = 20.8 \, \mathrm{J \, mol^{-1} \, K^{-1}}$. The specific heat capacity is therefore $10.4 \, \mathrm{kJ \, kg^{-1} \, K^{-1}}$.

A graph of the actual specific heat capacity of hydrogen (figure 180) shows that these predictions are essentially correct. The value of C_v begins to increase from $6.3 \, \mathrm{kJ \, kg^{-1} \, K^{-1}}$ at about 50 K, and reaches $10.4 \, \mathrm{kJ \, kg^{-1} \, K^{-1}}$ at about 300 K. Towards 1000 K we can begin to see evidence of the vibrational mode being excited.

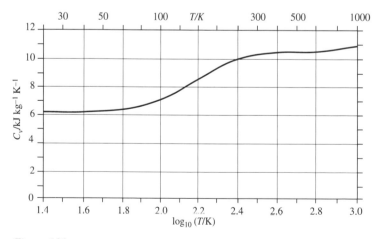

Figure 180

Problem 189

The speed of sound v in a gas is given by

$$v = \sqrt{\frac{\gamma kT}{m}},$$

where γ is the ratio of the principal heat capacities, k is Boltzmann's constant, T is the temperature and m is the mass of a molecule of the gas. At 15 °C the speed of sound in carbon dioxide (a linear molecule O–C–O) is found to be 264.7 m s^{-1}, while at 1000 °C it is 536.5 m s^{-1}. Interpret this observation in terms of quantum-mechanical effects on the heat capacity of a simple gas.

Solution

The relative molecular mass of CO_2 is $(12 + 2 \times 16) = 44$, so the molecular mass m is $44 \times 1.66 \times 10^{-27}$ kg $= 7.30 \times 10^{-26}$ kg. Thus at 15 °C $= 288$ K, the ratio of specific heats, γ, for CO_2 is

$$\gamma = \frac{v^2 m}{kT} = \frac{264.7^2 \times 7.30 \times 10^{-26}}{1.38 \times 10^{-23} \times 288} = 1.29.$$

Similarly, at 1000 °C $= 1273$ K, $\gamma = 1.20$.

We know that $\gamma = 1 + 2/F$ where F is the effective number of degrees of freedom, so at 288 K $F \approx 7$ and at 1273 K $F \approx 10$.

Now CO_2 is triatomic and linear, so it has three translational modes (as does any molecule), each of which contributes a single degree of freedom, and two rotational modes (rotation about the linear axis of the molecule not being permitted), each of which contributes a single degree of freedom. Since the total number of allowed modes is $3N$, where N is the number of atoms in the molecule, the number of vibrational modes is $9-3-2 = 4$, each of which contributes two degrees of freedom. Thus at sufficiently low temperatures, CO_2 will exhibit $F = 3$. At temperatures high enough to excite the rotational modes, but not high enough to excite the vibrational modes, $F = 5$, and the maximum number of degrees of freedom is $F = 13$. Thus the observations suggest that at 15 °C one of the four vibrational modes is excited, in addition to the translational modes and the two rotational modes, and that at 1000 °C two vibrational modes are excited, with the third partially excited.

[In fact, the excitation energies of the vibrational modes in the carbon dioxide molecule are 1.3, 1.3, 2.8, and 4.7, $\times 10^{-20}$ J, with corresponding temperatures of about 1000, 1000, 2000 and 3500 K respectively. It thus seems reasonable that at about 300 K the two lowest modes should each be half excited, whereas at 1300 K they are fully excited and the third mode is also half excited.]

Problem 190

According to simple kinetic theory the thermal conductivity of a gas is given by the expression

$$K = \tfrac{1}{3} c_v \lambda \langle c \rangle.$$

Explain the meaning of the symbols on the right-hand side of this equation and use it to show that the thermal conductivity of a gas should increase with temperature approximately, but not exactly, as $T^{1/2}$.

Make a reasoned estimate of the ratio of the thermal conductivity of methane gas to that of argon gas, at room temperature and atmospheric pressure, given the following information. Both molecules can be regarded as spheres, the radius of the methane molecule being 1.7 times the radius of the argon atom. The atomic weight of argon is 40; the molecular weight of methane is 16. Internal vibration of the methane molecule is not excited at room temperature.

Solution

λ is the mean free path of the gas molecules, i.e. the mean distance travelled by molecules between collisions with other molecules. $\langle c \rangle$ is their mean speed, and c_v is a heat capacity at constant volume. We can check which heat capacity it is by using dimensional analysis: The units of K are $\mathrm{W\,m^{-1}\,K^{-1}}$, and the units of $\lambda \langle c \rangle$ are $\mathrm{m^2\,s^{-1}}$, so the units of c_v must be $\mathrm{J\,m^{-3}\,K^{-1}}$. c_v is thus the heat capacity per unit volume at constant volume.

c_v is related to the molar heat capacity at constant volume, C_v, by

$$c_v = \frac{C_v n}{N_A},$$

where n is the number of molecules per unit volume and N_A is Avogadro's number. The mean free path λ is given approximately by

$$\lambda = \frac{1}{4\pi n r^2}$$

for a gas of spherical molecules of radius r, and the mean speed $\langle c \rangle$ is given by

$$\langle c \rangle = \sqrt{\frac{8RT}{\pi M}},$$

where R is the gas constant, T is the temperature and M is the molar mass. Putting these results together gives

$$K = \frac{C_v}{12\pi N_A r^2} \sqrt{\frac{8RT}{\pi M}}.$$

▶ Since C_v varies only very slowly with T (see problem 188), this implies that K should vary approximately as $T^{1/2}$, as required.

We can use this expression to write down the expected ratio of the thermal conductivity of methane to that of argon:

$$\frac{K_M}{K_A} = \frac{C_{v,M}}{C_{v,A}} \frac{r_A^2}{r_M^2} \sqrt{\frac{M_A}{M_M}}.$$

The molar heat capacity at constant volume, C_v, is $Rf/2$ where the gas molecules have f degrees of freedom. Since argon is a monatomic gas, $f = 3$ and $C_v = 3R/2$. The methane molecule is polyatomic and non-linear, so it is free to rotate about any axis, giving it an additional three degrees of freedom (we are told that none of the vibrational modes is excited). It thus has $f = 6$ and $C_v = 3R$. We can therefore put

$$\frac{K_M}{K_A} = 2\frac{1}{1.7^2} \sqrt{\frac{40}{16}} = 1.1.$$

▶ I.e. we predict that the conductivity of methane should be about 1.1 times that of argon. [In fact, the ratio is about 1.9, so our assumptions must be invalid.]

Problem 191

The equilibrium vapour pressures as a function of temperature for carbon tetrachloride (CCl_4) and mercury (Hg) are given in Table 9.

Table 9

	Vapour pressure in mm mercury	
$T/°C$ CCl_4		Hg
0 33		1.85×10^{-4}
20 91		1.20×10^{-3}
40 216		6.08×10^{-3}
60 451		2.52×10^{-2}
80 843		8.88×10^{-2}
100 1463		2.73×10^{-1}

What can you deduce from a plot of these measurements?

Solution

The variation of vapour pressure p_v with temperature T is given by the Clausius–Clapeyron equation

$$\frac{dp_v}{dT} = \frac{L}{T(V_v - V_1)},$$

where L is the molar latent heat of vaporisation, V_v is the molar volume of the vapour, and V_1 is the molar volume of the liquid.

If the vapour is an ideal gas, we may put

$$p_v V_v = RT.$$

Since $V_v \gg V_1$, we may ignore V_1, so that on combining these results we have

$$\frac{dp_v}{p_v} = \frac{L}{R} \frac{dT}{T^2}.$$

This can be integrated to give

$$\ln p_v = -\frac{L}{RT} + \text{constant}.$$

Thus a graph of $\ln p_v$ against $1/T$ should be linear, with a slope of $-L/R$. Also, by extrapolating or interpolating the graph to find the temperature at which p_v reaches atmospheric pressure (760 mm mercury), we can estimate the boiling point of the liquid.

The graphs (figure 181) show that both sets of data are described very well by the equation. If we perform a least-squares regression analysis on

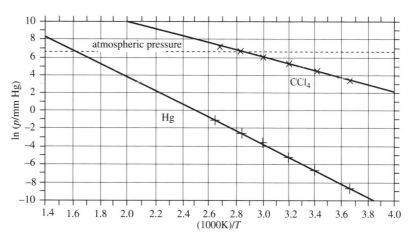

Figure 181

the CCl_4 data we find that the slope of the graph of $\ln p_v$ against $1/T$ is $(-3.87 \pm 0.04) \times 10^3$ K, so that the latent heat of vaporisation is
▶ 32.2 ± 0.3 kJ mol^{-1}. Interpolating the graph to find the value of $1/T$ at which the pressure is 760 mm Hg (i.e. $\ln p = 6.63$) gives
▶ $1/T = 2.859 \times 10^{-3}$ K^{-1}, so $T_b = 350$ K.
Similarly for mercury, the slope of the graph is $(-7.43 \pm 0.01) \times 10^3$ K,
▶ which gives 61.7 ± 0.1 kJ mol^{-1} for the latent heat of vaporisation. Extrapolation of the graph to $\ln p = 6.63$ gives $1/T = 1.613 \times 10^{-3}$ K^{-1},
▶ so $T_b = 620$ K.
The accepted values for CCl_4 are $L = 29$ kJ mol^{-1} and $T_b = 350$ K, and for Hg they are $L = 58$ kJ mol^{-1} and $T_b = 630$ K, so our calculated values are correct to within about 10%.

Problem 192

A mass of 1 kg is placed upon a block of melting ice. The weight bears upon an area of 1 mm^2. By how much must the temperature of the ice be lowered for it to resist penetration by the mass? Assume that no heat flows from the mass. ($g = 9.81$ m s^{-2}. Latent heat of fusion of ice $= 333$ kJ kg^{-1}. In cold water ice floats eleven-twelfths submerged.)

Solution

We can use the Clausius–Clapeyron equation to calculate the effect of pressure on the melting temperature of ice:

$$\frac{dp}{dT} = \frac{L}{(v_w - v_i)T}.$$

If the latent heat of fusion L is given as energy per unit mass (as it is in the problem), v_i and v_w must be the volumes per unit mass of ice and water respectively. These are clearly the reciprocals of the respective densities.

The problem specifies the density ρ_i of ice in relation to the density ρ_w of water through the statement that ice floats eleven-twelfths submerged. If we call this fraction f, we can use Archimedes' principle to calculate ρ_i as shown in figure 182.

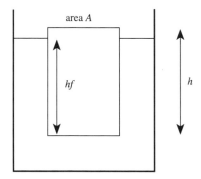

Figure 182

Let us consider a block of ice of cross-sectional area A and height h floating in water. The weight of the block is $\rho_i gAh$. A height hf of the block is submerged, so the volume of water displaced is Ahf and the weight of this water is $\rho_w gAhf$. This is the upthrust which balances the weight, so we can write $\rho_i gAh = \rho_w gAhf$, or

$$\rho_i = f\rho_w.$$

Thus

$$v_w - v_i = \frac{1}{\rho_w} - \frac{1}{f\rho_w} = \frac{1 - \dfrac{1}{f}}{\rho_w}.$$

Substituting this expression into the Clausius–Clapeyron equation gives

$$\frac{dp}{dT} = \frac{L\rho_w}{\left(1 - \dfrac{1}{f}\right)T},$$

and using the values quoted for L and f, and taking $T = 273$ K and $\rho_w = 1000$ kg m^{-3}, gives $dp/dT = -1.34 \times 10^7$ Pa K^{-1}. Since this is negative, the melting point of ice must be decreased by increasing the pressure, and the temperature of the ice must therefore be lowered to this value to prevent the mass from penetrating the ice. A mass of 1 kg bearing on an area of 1 mm^2 exerts a pressure of $9.81 \times 1/10^{-6}$ Pa $= 9.81 \times 10^6$ Pa, so the depression of the melting point is $(9.81 \times 10^6)/(1.34 \times 10^7)$ K $= 0.73$ K.

Problem 193

Water boils at 100 °C at sea level and at 83 °C on a mountain 5 km in height. Estimate the latent heat of vaporisation per mole of water. (For simplicity assume that air consists of a single molecular species of relative molecular mass 28.8, and take the ambient temperature to be 10 °C.)

Solution

Water boils when it reaches the temperature at which its saturated vapour pressure p_v is equal to the external (atmospheric) pressure. We expect the saturated vapour pressure to vary with temperature T as

$$p_v \propto \exp(-L/RT),$$

where L is the molar latent heat of vaporisation.

We also expect the atmospheric pressure p to vary with height z. Since the potential energy of a mole of air at height z is Mgz where M is the molar mass and g is the acceleration due to gravity, we may put

$$p \propto \exp(-Mgz/RT_a)$$

if the gravitational acceleration g and the ambient temperature T_a may be assumed to be constant (e.g. problem 184).

Thus if we put T_0 for the boiling point of water at sea level and T_z for the boiling point at height z, and p_0 for atmospheric pressure at sea level, we have two expressions for the vapour pressure at height z:

$$p_v = \frac{p_0 \exp(-L/RT_z)}{\exp(-L/RT_0)} = p_0 \exp(-Mgz/RT_a).$$

By eliminating p_0 and taking logarithms to base e, this may be rewritten as

$$\frac{L}{RT_0} - \frac{L}{RT_z} = -\frac{Mgz}{RT_a}.$$

Rearranging,

$$L = \frac{Mgz}{T_a}\left(\frac{1}{T_z} - \frac{1}{T_0}\right)^{-1}.$$

Substituting the values $M = 28.8 \times 10^{-3}$ kg, $g = 9.8$ m s^{-2}, $z = 5000$ m, $T_a = 283$ K, $T_0 = 373$ K and $T_z = 356$ K gives $L = 39$ kJ mol^{-1}. (The accepted value is about 41 kJ mol^{-1}, so our simple method has underestimated L by about 5%.)

Problem 194

A gas obeying

$$p(V - b) = RT$$

has a heat capacity per mole, at constant volume, of C_v.

(a) Write down Maxwell's relations, and use them to show that the internal energy U is a function only of T by finding $(\partial U/\partial T)_V$, $(\partial U/\partial V)_T$ and hence also $(\partial U/\partial p)_T$.

(b) Show that for an adiabatic expansion $p(V - b)^\gamma = $ constant, where γ is the ratio of principal heat capacities and $C_p - C_V = R$.

Solution

(a) The first result, $(\partial U/\partial T)_V$, is by definition equal to C_v.

The four Maxwell relations can be written as

$$\left(\frac{\partial T}{\partial V}\right)_S = -\left(\frac{\partial p}{\partial S}\right)_V, \quad \left(\frac{\partial T}{\partial p}\right)_S = \left(\frac{\partial V}{\partial S}\right)_p,$$

$$\left(\frac{\partial S}{\partial V}\right)_T = \left(\frac{\partial p}{\partial T}\right)_V, \quad \left(\frac{\partial S}{\partial p}\right)_T = -\left(\frac{\partial V}{\partial T}\right)_p.$$

The first law of thermodynamics can be written as

$$dU = T\,dS - p\,dV,$$

so if we perform the differentiation with respect to V at constant T we obtain

$$\left(\frac{\partial U}{\partial V}\right)_T = T\left(\frac{\partial S}{\partial V}\right)_T - p.$$

Substituting the appropriate Maxwell relation gives

$$\left(\frac{\partial U}{\partial V}\right)_T = T\left(\frac{\partial p}{\partial T}\right)_V - p.$$

Now for the gas defined in the problem, we have

$$p = \frac{RT}{V - b},$$

so that

$$\left(\frac{\partial p}{\partial T}\right)_V = \frac{R}{V - b} = \frac{p}{T}.$$

Thus, for this gas,

▶ $$\left(\frac{\partial U}{\partial V}\right)_T = 0.$$

Similarly, beginning with $dU = T\,dS - p\,dV$ and performing the differentiation with respect to p at constant T gives

$$\left(\frac{\partial U}{\partial p}\right)_T = T\left(\frac{\partial S}{\partial p}\right)_T - p\left(\frac{\partial V}{\partial p}\right)_T.$$

Substitution of the appropriate Maxwell relation gives

$$\left(\frac{\partial U}{\partial p}\right)_T = -T\left(\frac{\partial V}{\partial T}\right)_p - p\left(\frac{\partial V}{\partial p}\right)_T.$$

From the equation of state we have

$$\left(\frac{\partial V}{\partial T}\right)_p = \frac{R}{p} \text{ and } \left(\frac{\partial V}{\partial p}\right)_T = -\frac{RT}{p^2},$$

so that

▶ $$\left(\frac{\partial U}{\partial p}\right)_T = 0.$$

Since we have shown that U depends on neither p nor V if the temperature is held constant, it follows that U is a function only of T.

(b) For an adiabatic expansion, $dS = 0$ so $dU = -p\,dV$. However, since we have just shown that U depends only on T we may also put $dU = C_v dT$, and thus

$$C_v dT + p\,dV = 0.$$

We can substitute from the equation of state $p(V - b) = RT$ to eliminate p and obtain an expression involving only V and T:

$$C_v dT + \frac{RT\,dV}{V - b} = 0.$$

This differential equation can be solved by dividing through by T and then integrating:

$$\int \frac{C_v dT}{T} + \int \frac{R\,dV}{V - b} = \text{constant},$$

which gives

$$C_v \ln T + R \ln (V - b) = \text{constant}.$$

Substitution from the equation of state to eliminate T gives

$$C_v \ln p + (R + C_v) \ln (V - b) = \text{constant}.$$

Recalling that $R + C_v = C_p$, and dividing this expression by C_v, gives

$$\ln p + \gamma \ln (V - b) = \text{constant}.$$

Finally, taking the antilogarithm of this expression gives the required result:

▶ $$p(V - b)^\gamma = \text{constant}.$$

Problem 195

A body has a constant heat capacity C_p and an initial temperature T_1. It is placed in contact with a heat reservoir at temperature T_2 and comes into equilibrium with it at constant pressure. Assuming T_2 is greater than T_1, calculate the entropy change of the universe and show that this is always positive.

Solution

Since T_2 is greater than T_1, heat flows from the reservoir to the body. The reservoir may be assumed to have infinite heat capacity, i.e. its temperature is unchanged, but the effect of heat flowing to the body will be to increase the body's temperature.

Consider a quantity of heat dQ flowing from the reservoir to the body. The entropy change of the body will be $+dQ/T$, where T is the body's temperature, and the entropy change of the reservoir will be $-dQ/T_2$.

The change in total entropy during this process will thus be

$$dQ\left(\frac{1}{T} - \frac{1}{T_2}\right),$$

and since T will always be less than or equal to T_2, this entropy change will always be greater than or equal to zero. Thus the entropy of the universe must increase as a result of this process.

We can relate dQ to the temperature change dT of the body through $dQ = C_p dT$, so

$$dS_{\text{univ}} = C_p\frac{dT}{T} - C_p\frac{dT}{T_2}.$$

The total entropy change of the universe is found by integrating this expression between the limits $T = T_1$ to $T = T_2$:

$$\Delta S_{\text{univ}} = C_p\int_{T_1}^{T_2}\frac{dT}{T} - \frac{C_p}{T_2}\int_{T_1}^{T_2}dT$$

$$= C_p\left(\ln\left[\frac{T_2}{T_1}\right] - \left[1 - \frac{T_1}{T_2}\right]\right).$$

We can show mathematically (as opposed to physically) that this must be greater than zero by putting the dimensionless variable $x = 1 - T_1/T_2$, and noting that, since T_2 is greater than T_1, x must be positive. Our expression for ΔS_{univ} can be rewritten in terms of x as

$$\Delta S_{\text{univ}}/C_p = -\ln(1-x) - x$$
$$= x + x^2/2 + x^3/3 + x^4/4 + \ldots -x,$$

which is clearly greater than zero if x is greater than zero.

Problem 196

A cold body of heat capacity A and temperature T_1 is connected to the environment (at constant temperature T_2, with $T_2 > T_1$) by an ideal heat engine. Find the maximum work W that can be extracted from the system.

This work is then returned to the heat engine so as to heat the body. Show that the highest temperature T_3 that can be reached is given by

$$T_3 - T_2\left(1 + \ln\frac{T_3}{T_2}\right) = \frac{W}{A}.$$

Solution

Since the body does not remain at temperature T_1, let us write T for its general temperature. We can draw a diagram to illustrate the first process, as shown in figure 183.

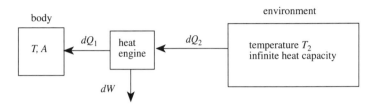

Figure 183

The first law of thermodynamics (conservation of energy) shows that

$$dQ_2 = dQ_1 + dW,$$

and the second law shows that, for an ideal heat engine (for which dW will be maximum),

$$\frac{dQ_1}{T} = \frac{dQ_2}{T_2}.$$

Combining these, we find that

$$dW = \left(\frac{T_2}{T} - 1\right)dQ_1.$$

We also know that $dQ_1 = A\,dT$, so we can write

$$dW = \left(\frac{T_2}{T} - 1\right)A\,dT.$$

We can now integrate this expression. The left-hand side will give the maximum work W available from the system when we integrate the right-hand side from $T = T_1$ to $T = T_2$:

$$W = A\int_{T_1}^{T_2}\left(\frac{T_2}{T} - 1\right)dT = A[T_2 \ln T - T]_{T_1}^{T_2},$$

which gives

$$W = A\left(T_2 \ln\frac{T_2}{T_1} - T_2 + T_1\right).$$

Now we use this work (which we have stored somewhere) to drive the heat engine in reverse in order to raise the temperature of the body above T_2, as shown in figure 184.

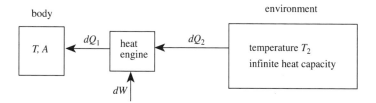

Figure 184

This time, application of the first law gives

$$dQ_1 = dQ_2 + dW,$$

but the second law gives the same condition as before. Combining the two conditions as before, and writing $dQ_1 = A\,dT$, we obtain

$$A\,dT\left(1 - \frac{T_2}{T}\right) = dW.$$

This expression can be integrated, with W running from zero to the value we calculated in the earlier part of the problem, and T running from T_2 to T_3:

$$\int_{T_2}^{T_3}\left(1 - \frac{T_2}{T}\right)dT = \int_0^W \frac{dW}{A},$$

which gives

$$[T - T_2\ln T]_{T_2}^{T_3} = \frac{W}{A},$$

$$T_3 - T_2 - T_2\ln\frac{T_3}{T_2} = \frac{W}{A},$$

which can finally be rearranged to give the required form

▶ $$T_3 - T_2\left(1 + \ln\frac{T_3}{T_2}\right) = \frac{W}{A}.$$

[Although the problem does not ask for it, we can substitute into this expression our earlier expression for W to obtain the relationship between T_1, T_2 and T_3. This gives

$$T_3 - T_2\left(1 + \ln\frac{T_3}{T_2}\right) = T_2\ln\frac{T_2}{T_1} - T_2 + T_1,$$

which can be simplified to

$$T_3 - T_1 = T_2\left(\ln\frac{T_3}{T_2} + \ln\frac{T_2}{T_1}\right) = T_2\ln\frac{T_3}{T_1}.$$

As an example, let us suppose that $T_1 = 273$ K and $T_2 = 300$ K. The value of T_3 in kelvins must then satisfy the equation $300\ln T_3 - T_3 = 300\ln 273 - 273 = 1409.8$. This cannot be solved analytically, but a numerical or graphical solution gives $T_3 \approx 329$. Thus a body initially at $0\,°C$ can be raised to a temperature of $56\,°C$ using an ideal heat engine operating between the body and its environment if the environment is at $27\,°C$.]

Problem 197

Show that during a reversible isothermal expansion of an ideal gas from volume V_1 to volume V_2 there is a change of entropy

$$\Delta S = \int_{V_1}^{V_2}\left(\frac{\partial p}{\partial T}\right)_V dV.$$

If the volume of 2 moles of an ideal gas is doubled during such an expansion, calculate the entropy change.

Solution

We are concerned with the change in entropy for a change in volume at constant temperature, so we need to consider the expression

$$\left(\frac{\partial S}{\partial V}\right)_T.$$

Now we know from Maxwell's thermodynamic relations that this is equal to

$$\left(\frac{\partial p}{\partial T}\right)_V,$$

so it follows directly that

$$\Delta S = \int_{V_1}^{V_2}\left(\frac{\partial p}{\partial T}\right)_V dV.$$

For N moles of an ideal gas, $pV = NRT$ (where R is the gas constant), so

$$\left(\frac{\partial p}{\partial T}\right)_V = \frac{NR}{V}.$$

Thus

$$\Delta S = NR \int_{V_1}^{V_2} \frac{dV}{V} = NR \ln \frac{V_2}{V_1}.$$

▶ Taking $N = 2$, $R = 8.314 \, \mathrm{J \, K^{-1} \, mol^{-1}}$ and $V_2/V_1 = 2$ gives $\Delta S = 11.5 \, \mathrm{J \, K^{-1}}$.

Problem 198

Why does an ideal gas cool when it expands adiabatically and reversibly?

One mole of an ideal gas at 300 K expands isothermally from a pressure of $2 \times 10^6 \, \mathrm{N \, m^{-2}}$ to a pressure $2 \times 10^5 \, \mathrm{N \, m^{-2}}$. Calculate the heat absorbed by the gas, the change in internal energy and the change in entropy in the following cases:

(a) the gas expands reversibly;
(b) the gas expands against zero pressure.

Solution

The first law of thermodynamics can be expressed as

$$\Delta U = dq + dw,$$

where ΔU is the change in internal energy of a system, dq is the heat added to the system, and dw is the work done on the system. Now for an adiabatic change, $dq = 0$, and for a reversible expansion, $dw < 0$ (since the gas expands against an external pressure), so $\Delta U < 0$. For an ideal gas the internal energy U is proportional to the temperature, so $dT < 0$.

▶
▶   (a) During an isothermal expansion of an ideal gas, $\Delta T = 0$ so $\Delta U = 0$. The work done on the gas is given by

$$dw = -p \, dV$$

and

$$pV = NRT,$$

where N is the number of moles of gas, so

$$dV = -\frac{NRT}{p^2}dp.$$

Thus the total work done on the gas during an expansion from pressure p_0 to pressure p_1 is

$$dw = NRT\int_{p_0}^{p_1}\frac{dp}{p} = NRT\ln\frac{p_1}{p_0}.$$

Taking $N = 1$, $T = 300$ K and $p_1/p_0 = 1/10$ gives $dw = -5.74$ kJ. Since
▶ $\Delta U = dq + dw$ and $\Delta U = 0$, the heat input to the gas must be
▶ $dq = +5.74$ kJ.

This heat input is added at constant temperature, so the entropy change
▶ of the gas is given by dq/T, i.e. $\Delta S = +19.1$ J K^{-1}. The entropy change of the universe is zero since the process is reversible.

(b) If the expansion is performed against zero pressure, the process is irreversible. Since the temperature of the gas does not change, we must
▶ still have $\Delta U = 0$. However, at zero pressure, the surroundings cannot do
▶ any work on the gas, so we must also have $dq = 0$. Since entropy is a state variable (depends only on the final state of the gas, not the route used to achieve this state), the entropy change of the gas must be the same as
▶ calculated in part (a), i.e. $\Delta S = +19.1$ J K^{-1}. The entropy change of the surroundings must be zero, so the total entropy change of the universe is $+19.1$ J K^{-1}, i.e. positive, which is what we expect for an irreversible change.

Problem 199

1 kg of ice at 0 °C floats in 5 kg of water at 50 °C, the whole being thermally isolated. When thermal equilibrium has been reached, by how much will the entropy of the system have changed? (Specific heat of water $= 4200$ J kg^{-1} K^{-1}. Latent heat of fusion of ice $= 333$ kJ kg^{-1}.)

Solution

The ice will remain at 0 °C until it has all been melted. The total heat input required to do this is clearly

$$M_i L,$$

where M_i is the mass of the ice and L is its latent heat of fusion. The entropy of the ice thus increases by

$$\frac{M_i L}{T_i},$$

where T_i is its temperature. Substituting the given values, we find that the entropy of the ice increases by $1219.1\ \mathrm{J\,K^{-1}}$ during melting.

During the same period, the water loses entropy as it gives up heat to the ice. The calculation is more complicated because the temperature of the water does not remain constant (figure 185).

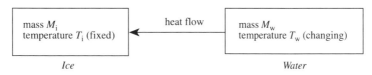

| mass M_i temperature T_i (fixed) | ← heat flow | mass M_w temperature T_w (changing) |

Ice   *Water*

Figure 185

If heat dQ is *added* to the water, its temperature will increase by

$$dT = \frac{dQ}{M_w C},$$

where M_w is the mass of water and C is its specific heat capacity. Its entropy will thus increase by

$$dS = \frac{dQ}{T} = M_w C \frac{dT}{T}.$$

We can integrate this expression to find the total entropy change ΔS when the mass M_w of water changes its temperature from T_0 to T_1. It gives

$$\Delta S = M_w C \ln \frac{T_1}{T_0}.$$

Now during the melting of the ice, heat $M_i L$ is transferred from the water to the ice, so the temperature of the water must decrease by

$$\frac{M_i L}{M_w C}.$$

Substituting the values given, we find that the temperature of the water will thus decrease by $15.86\ \mathrm{K}$ so that its final temperature will be $307.29\ \mathrm{K}$ compared with an initial temperature of $323.15\ \mathrm{K}$. The entropy change of the water is thus

$5 \times 4200 \times \ln{(307.29/323.15)} \text{ J K}^{-1}$
$= -1056.8 \text{ J K}^{-1}.$

The total change of entropy of the system in melting the ice is therefore $1219.1 - 1056.8 = 162.3 \text{ J K}^{-1}$. However, the system is not yet in equilibrium because it now consists of 5 kg of water at 307.29 K and 1 kg of water at 273.15 K. Heat will be transferred from the 5 kg of water to the 1 kg of water until their temperatures are equal (figure 186).

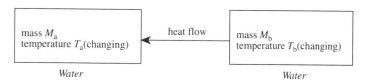

Figure 186

If we call the two masses of water M_a and M_b and their initial temperatures T_a and T_b, and let the final (equilibrium) temperature be T, the heat gained by M_a must equal the heat lost by M_b, thus

$$CM_a(T - T_a) = CM_b(T_b - T).$$

Taking $T_a = 273.15 \text{ K}$, $T_b = 307.29 \text{ K}$, $M_a = 1 \text{ kg}$, $M_b = 5 \text{ kg}$, we can solve for the equilibrium temperature T to find

$$T = 301.60 \text{ K}$$

(which is 28.5 °C). Using the expression for ΔS which we derived earlier, the change in the entropy of the mass M_a is thus

$1 \times 4200 \times \ln{(301.60/273.15)} \text{ J K}^{-1}$
$= 416.1 \text{ J K}^{-1},$

and the change in the entropy of the mass M_b is

$5 \times 4200 \times \ln{(301.60/307.29)} \text{ J K}^{-1}$
$= -392.5 \text{ J K}^{-1}.$

The total change in entropy of the system in coming to equilibrium from its initial state is thus

$162.3 + 416.1 - 392.5 \text{ J K}^{-1}$
$= 186 \text{ J K}^{-1}.$

[There is another possible approach to this problem. Since both the internal energy E and the entropy S are state variables, we can calculate

E and S for ice and for water. We can define the zero values of E and S arbitrarily, since we are only interested in changes in these quantities. For convenience, we will take the zero values of each to refer to ice at its melting point T_m. Let us first consider the energy content of a mass m of water at temperature T. In order to melt a mass m of ice, we must add heat Lm, and in order to raise the temperature to T we must add heat $mc(T - T_m)$. No work is done on the system, so we have

$$E/m = L + CT - CT_m.$$

Now let us consider the entropy content. In order to melt a mass m of ice, we must add the heat Lm at temperature T_m, so the entropy increases by Lm/T_m. The addition of further heat dQ will raise the temperature by $dT = dQ/Cm$, so the entropy increase dS can be written as $dQ/T = CmdT/T$. Thus the increase in entropy in raising the temperature from T_m to T is $Cm \ln(T/T_m)$, and hence

$$S/m = L/T_m + C \ln(T/T_m).$$

Initially, the system consists of 1 kg of ice at $T_m = 273.15$ K, which has by definition zero energy and zero entropy, and 5 kg of water at 323.15 K. The total energy of the system is thus

$$5(333\,000 + 4200 \times 323.15 - 4200 \times 273.15)\ \text{J} = 2\,715\,000\ \text{J}$$

and the total entropy is

$$5(333\,000/273.15 + 4200 \ln(323.15/273.15))\ \text{J K}^{-1}$$
$$= 9625.558\ \text{J K}^{-1}.$$

When the system has reached equilibrium, it consists of 6 kg of water at temperature T. Since the total energy is conserved, we must have $E/m = 2\,715\,000/6\ \text{J kg}^{-1} = 452\,500\ \text{J kg}^{-1} = L + CT - CT_m$, so

$$T = \frac{452\,500 + 4200 \times 273.15 - 333\,000}{4200}\ \text{K} = 301.60\ \text{K}.$$

The total entropy of the system is thus

$$6(333\,000/273.15 + 4200 \ln(301.60/273.15))\ \text{J K}^{-1} = 9811.492\ \text{J K}^{-1}$$

and the entropy increase of the system is

$$9811.492 - 9625.558\ \text{J K}^{-1} = 186\ \text{J K}^{-1}$$

as before.]

Problem 200

A house has heat capacity C_H and loses heat to the surroundings at a rate $A(T_H - T_S)$, where A is a constant and T_H and T_S are the temperatures of the house and surroundings, respectively. An electrically powered heat pump with an ideal performance takes heat from a large reservoir at temperature T_S and delivers heat to radiators in the house at a constant temperature T_R where $T_R > T_S$. The radiators provide heat to the house at a rate $B(T_R - T_H)$ where B is a constant. The house is initially at temperature T_S.

(a) What is the efficiency of the heat pump?

(b) What is the initial power consumption of the heat pump?

(c) What is the initial warming rate of the house?

(d) If the heating is left on for a long time, what is the equilibrium temperature in the house?

(e) Sketch a graph of T_H versus time.

(f) Show that the power consumption at the equilibrium temperature is

$$\frac{AB(T_R - T_S)^2}{(A + B)T_R}.$$

Solution

It will be helpful to draw a diagram (figure 187) showing the energy paths involved in this problem.

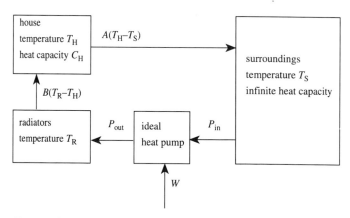

Figure 187

(a) The heat pump accepts heat at a rate P_{in} from the surroundings, and rejects it at a rate P_{out} to the radiators. Work is input to the pump at a rate W, so by conservation of energy

$$P_{out} = P_{in} + W.$$

For an ideal heat pump,

$$\frac{P_{out}}{T_R} = \frac{P_{in}}{T_S}.$$

Combining these two results gives

$$P_{out}(1 - T_S/T_R) = W,$$

so the efficiency η of the heat pump, defined as P_{out}/W, is

▶
$$\eta = \frac{1}{1 - T_S/T_R}.$$

(b) Initially the house is at temperature T_S, so the rate at which heat is supplied to the house by the radiators must be $B(T_R - T_S)$. This heat is supplied by the heat pump, so $P_{out} = B(T_R - T_S)$ and the power consumption of the heat pump, W, is given by $W = P_{out}/\eta$, so

▶
$$W = B(T_R - T_S)(1 - T_S/T_R) = \frac{B(T_R - T_S)^2}{T_R}.$$

(c) Initially heat is supplied to the house at a rate $B(T_R - T_S)$ and dissipated at a rate of zero (since $T_H = T_S$). The heat capacity of the house is C_H, so the initial rate of warming must be

▶
$$\frac{dT_H}{dt} = \frac{B(T_R - T_S)}{C_H}.$$

(d) At equilibrium, the rate $B(T_R - T_H)$ at which heat is supplied to the house must be equal to the rate $A(T_H - T_S)$ at which heat is dissipated. Equating these terms, and rearranging to make T_H the subject, gives

▶
$$T_H = \frac{AT_S + BT_R}{A + B}.$$

(e) For some value of T_H intermediate between the initial value T_S and the final (equilibrium) value calculated in part (d), the net rate at which heat is being transferred to the house is $B(T_R - T_H) - A(T_H - T_S)$, so that

$$\frac{dT_H}{dt} = \frac{1}{C_H}(B[T_R - T_H] - A[T_H - T_S]).$$

If we put $T'_H = T_H - T_H(\text{equilib})$, the differential equation can be simplified to

$$\frac{dT'_H}{dt} = -\frac{(A + B)}{C_H} T'_H.$$

The solution of this equation is an exponential decay with a time constant of $C_H/(A + B)$, so a suitable sketch of T_H versus time would be as shown in figure 188.

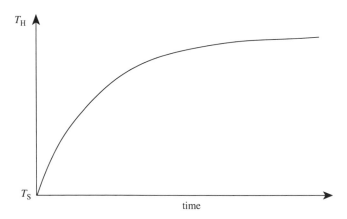

Figure 188

   (f) At equilibrium, the heat supplied by the heat pump must be equal to $B(T_R - T_H)$, where T_H is the equilibrium temperature calculated in part (d). Substituting our expression from part (d) gives

$$\frac{AB(T_R - T_S)}{A + B}.$$

The power consumption is found by dividing this value by the efficiency of the heat pump:

$$W = \frac{AB}{A + B}(T_R - T_S)(1 - T_S/T_R) = \frac{AB}{A + B}\frac{(T_R - T_S)^2}{T_R}$$

as required.

Index

The numbers refer to the problems.